Springer Complexity

Springer Complexity is an interdisciplinary program publishing the best research and academic-level teaching on both fundamental and applied aspects of complex systems – cutting across all traditional disciplines of the natural and life sciences, engineering, economics, medicine, neuroscience, social and computer science.

Complex Systems are systems that comprise many interacting parts with the ability to generate a new quality of macroscopic collective behavior the manifestations of which are the spontaneous formation of distinctive temporal, spatial or functional structures. Models of such systems can be successfully mapped onto quite diverse "real-life" situations like the climate, the coherent emission of light from lasers, chemical reaction-diffusion systems, biological cellular networks, the dynamics of stock markets and of the internet, earthquake statistics and prediction, freeway traffic, the human brain, or the formation of opinions in social systems, to name just some of the popular applications.

Although their scope and methodologies overlap somewhat, one can distinguish the following main concepts and tools: self-organization, nonlinear dynamics, synergetics, turbulence, dynamical systems, catastrophes, instabilities, stochastic processes, chaos, graphs and networks, cellular automata, adaptive systems, genetic algorithms and computational intelligence.

The three major book publication platforms of the Springer Complexity program are the monograph series "Understanding Complex Systems" focusing on the various applications of complexity, the "Springer Series in Synergetics", which is devoted to the quantitative theoretical and methodological foundations, and the "SpringerBriefs in Complexity" which are concise and topical working reports, case-studies, surveys, essays and lecture notes of relevance to the field. In addition to the books in these two core series, the program also incorporates individual titles ranging from textbooks to major reference works.

Understanding Complex Systems

Founding Editor: J.A. Scott Kelso

Future scientific and technological developments in many fields will necessarily depend upon coming to grips with complex systems. Such systems are complex in both their composition – typically many different kinds of components interacting simultaneously and nonlinearly with each other and their environments on multiple levels – and in the rich diversity of behavior of which they are capable.

The Springer Series in Understanding Complex Systems series (UCS) promotes new strategies and paradigms for understanding and realizing applications of complex systems research in a wide variety of fields and endeavors. UCS is explicitly transdisciplinary. It has three main goals: First, to elaborate the concepts, methods and tools of complex systems at all levels of description and in all scientific fields, especially newly emerging areas within the life, social, behavioral, economic, neuro- and cognitive sciences (and derivatives thereof); second, to encourage novel applications of these ideas in various fields of engineering and computation such as robotics, nano-technology and informatics; third, to provide a single forum within which commonalities and differences in the workings of complex systems may be discerned, hence leading to deeper insight and understanding.

UCS will publish monographs, lecture notes and selected edited contributions aimed at communicating new findings to a large multidisciplinary audience.

For further volumes:
http://www.springer.com/series/5394

Andrzej Nowak • Katarzyna Winkowska-Nowak
David Brée

Editors

Complex Human Dynamics

From Mind to Societies

Springer

Editors
Andrzej Nowak
Department of Psychology
University of Warsaw
Warsaw
Poland

Katarzyna Winkowska-Nowak
Warsaw School of Social Sciences
and Humanities
Warsaw
Poland

David Brée
Institute for Social Studies
University of Warsaw
Warsaw
Poland

ISSN 1860-0832   ISSN 1860-0840 (electronic)
ISBN 978-3-642-31435-3 ISBN 978-3-642-31436-0 (eBook)
DOI 10.1007/978-3-642-31436-0
Springer Heidelberg New York Dordrecht London

Library of Congress Control Number: 2012951923

Printed on acid-free paper

Springer is part of Springer Science+Business Media (www.springer.com)

Preface

Arguably, there is nothing more complex than the human mind. The myriad of thoughts and feelings, constantly evolving and reconfiguring into new meanings, has no parallel in natural phenomena. The complexity of human experience is even more enhanced in social contexts, where social and cultural elements are created and undergo change as individuals interact. The social sciences, which focus on human experience and behavior, have dealt with the complexity of human and social phenomena by developing qualitative theories focused more on understanding than prediction. Precise, quantitative models originating in traditional physics were not suited to grasping the dynamics of psychological and social phenomena.

This has changed dramatically as recent advances in the natural sciences have enhanced the capacity of the natural sciences to study not only physical and biological systems but social systems as well. The rise of complexity science can be traced to the development of new theories in natural sciences in the early 1980s. Such themes as deterministic chaos, artificial neural networks, genetic algorithms, emergence, and self-organization have not only profoundly transformed the natural sciences but also have captured the attention of the popular press and thus the general public.

Currently a revolution is occurring in the social sciences. The current advances in understanding how complex systems work has made psychological and social phenomena amenable to analysis with the tools of the natural sciences. Precise formalisms, including networks, agent-based models, and nonlinear dynamical systems, have started to transform our understanding of the psychological and social sciences. Social phenomena are increasingly at the center of the emerging interdisciplinary complexity science. As computer simulations became the primary tool of the complexity approach to social phenomena, models of social processes began to become precise and acquire predictive power.

The emerging complexity science is increasingly focused on social phenomena. Such issues as opinion dynamics and dynamics of social networks and financial markets are now at the center of interest of physicists and computer scientists. New technologies have provided unprecedented amounts of data, from the Internet, mobile phones, and sensors, concerning patterns of activity, locations, and movements, as well as communication between individuals. Complexity scientists are developing new, quantitative theories of social processes.

The new, complexity-based social theories have very limited awareness of the relevant knowledge existing in the social sciences. Correspondingly, the social sciences have only recently started to be interested in the contributions of complexity science to social phenomena. As a result, the two approaches to studying social phenomena are developing in large part on parallel tracks. The vast majority of research that takes the complexity approach to social phenomena comes from the formal sciences and follows their standards.

This book presents theories and research conducted in psychology that fall into the domain of complexity science, but from the perspective of the social sciences. It contains both reviews of the relevant literature and original empirical results. It discusses psychological and social phenomena in a way that makes them amenable to analysis by the tools of the natural sciences and shows examples of the use of these tools. It goes beyond the social topics most often discussed in the complexity approach to the social sciences, such as opinion dynamics. It concentrates on such issues as emergence of brain functions, detection of patterns in cognitive systems, dynamics of interpersonal attraction, dynamics of involvement in sport, group dynamics, and dynamics of conflict.

It shows how phenomena occurring at different levels of psychological and social reality, from brain to society, can be understood from the perspective of complex systems. On the one hand, processes occurring at all levels can be seen in a similar way, as produced by elements influencing each other in time. Whether the elements are neurons, thoughts, individuals, or social groups, the processes at each level are revealed as each element changes in time, responding to multiple influences of other connected elements of the system. Higher-level phenomena emerge from the interaction of lower-level elements. Neurons are interconnected by synapses and thoughts emerge as neurons change their frequency and pattern of firing in response to influences of other neurons. Cognitive elements are interrelated by functional and associative links and their interactions emerge as cognitive functions, such as pattern recognition, language, and consciousness. Social relations interconnect individuals in social networks. Processes occurring in these networks, such as social influence, flow of information, and social interdependence, produce dynamics of dyads, groups, and societies. The pattern of these influences determines the dynamical properties of emergent phenomena at the higher level. Synchronization of neuronal activations underlies recognition. Patterns of interaction of thoughts, feelings, and behaviors determine sport involvement. Patterns of interaction between individuals or groups set the stage for escalation or de-escalation in interpersonal and societal conflicts.

On the other hand, each level of social reality is unique, with vastly different elements and very different relations between elements and different ways in which the elements influence each other. The specific nature of the elements and their relations defines the content, the psychological as well as the social nature of the processes occurring in the mind and in society. This aspect can be understood with the use of the insights of psychological and social theories. Although the dynamics of self-esteem, involvement in sports, and conflict may be understood in common terms of attractor dynamics, the nature of each of these three phenomena is completely different and without appreciating these differences, constructing meaningful theory of each of them is impossible.

This book is intended to bridge the gap between the understanding of social processes from both the perspective of the social and the natural sciences. For social scientists, it highlights insights in the understating of psychological and social processes that can be achieved by adopting the perspective of complexity science. For the readers with background in the formal sciences, it presents relevant theories of the social sciences and highlights the way they can be amenable to analysis with the tools of the formal sciences. It also shows which of the tools developed in the formal sciences have proven to be more useful for investigating social phenomena.

This book describes psychological and social processes occurring at different levels of reality. It starts from the level of the brain then progresses through the levels of an individual, the dyad, the social group and ends up at the societal level. On each level, it concentrates on specific selected topics analyzed from a complexity perspective. It does not aim at a comprehensive description of all the phenomena occurring at each level that have been or could be analyzed with the tools of complexity science. We hope that the provision of specific in-depth examples of complexity-inspired social theories serves better the purpose of showing ways to apply complexity science to psychological and social dynamics than would a textbook-oriented approach.

The chapters of this book reflect lines of research conducted by the researchers in the Centre of Complex Systems of the Institute of Social Studies, at University of Warsaw. The majority of the researchers of this center, in contrast to most complexity-oriented research groups, have their background in psychology and sociology, some in computer science and biology, physics, engineering, and mathematics. The research group has a clearly interdisciplinary focus. It is involved in a lot of international interdisciplinary collaboration, mainly in the area of future and emerging technologies. In most cases, the members of the group collaborate with physicists and computer scientists providing expertise in the social area. The researchers from the Warsaw group combine in their research social theory and computer simulations, but also include empirical work with human participants.

This book presents simple models of complex psychological and social phenomena. The models presented follow the principle of *Dynamical minimalism* (Nowak 2004). In complex systems, even very simple rules of interaction between simple elements may lead to the emergence of complex properties at the level of the system. It follows that at least some complex phenomena may have very simple explanations. Computer simulations are used to explore which simple rules may explain the complex phenomena that have been observed. Simple elements must interact in time for complex phenomena to emerge. Minimalist theories based on emergence must be dynamic. In this approach, unlike approaches in the traditional social sciences, one tries to concentrate on the most essential properties of the phenomena to be explained, rather than trying to capture the phenomenon in its naturally occurring complexity. The task of the researcher is to try to find the set of the simplest possible, but realistic, rules that can reproduce the essence of the phenomena to be explained.

The chapters are arranged in the order dictated by levels of description: starting from the brain and ending at the societal level. Although the topics of the chapters are very diverse, several concepts coming from the formal sciences are central to our understanding of the psychological and social phenomena. The concept of attractor, for example, allows us to understand how stability can be combined with constant change in understanding the dynamics of attitudes, self-esteem involvement in sport, and conflict. Synchronization is central for our understanding of such diverse topics as how brain activity produces perceptions and thoughts, interpersonal attraction, and how individuals interact in groups. Networks provide formalisms to describe the functional connectivity in the brain, cognitive representations in the mind, and patterns of influences in social groups.

Reference

Nowak, A.: Dynamical minimalism: why less is more in psychology. Personal. Soc. Psychol. Rev. **8**, 183–192 (2004)

Contents

Chapter 1
Dynamical Social Psychology: An Introduction

Andrzej Nowak, Robin Vallacher, Urszula Strawińska, and David S. Brée

Abstract In this chapter we outline the background to dynamical social psychology as it stood before the research described in later chapters of this book. This background will help readers who are not familiar with either social psychology or complexity science to follow those chapters more easily. It will focus on two domains within dynamical social psychology: social influence on opinion formation and the concept of self. It will also consider three aspects of dynamical social psychology that set it apart from previous theories of social psychology: the effect of the degree of coherence between the elements of a system, which explains the different behaviors we see under different circumstances, how emotions regulate other psychological systems, and the drive to minimalism, by which behavior which appears to be complex may be understood from a model of the underlying elements interacting under simple rules.

A. Nowak (✉)
Department of Psychology, University of Warsaw, Stawki 5/7, 00183 Warsaw, Poland
e-mail: andrzejn232@gmail.com

R. Vallacher
Department of Psychology, Florida Atlantic University, Glades Rd. 777, Boca Raton, FL 33431, USA
e-mail: vallacher@fau.edu

U. Strawińska
Department of Psychology, Warsaw School of Social Sciences and Humanities, Chodakowska 19/31, 03815 Warsaw, Poland
e-mail: urszula.strawinska@swps.edu.pl

D.S. Brée
Institute for Social Studies, University of Warsaw, Stawki 5/7, 00183 Warsaw, Poland
e-mail: davidsbree@gmail.com

A. Nowak et al. (eds.), *Complex Human Dynamics*, Understanding Complex Systems, DOI 10.1007/978-3-642-31436-0_1, © Springer-Verlag Berlin Heidelberg 2013

Human social experience most probably is the most complex and dynamic phenomena subject to scientific scrutiny. Social psychologists have made ambitious attempts to capture the essence and detail of this experience. Occasionally they have formulated their insights into a precise and coherent theory, but such attempts have usually been abandoned as the next generation of experiments reveal intriguing social phenomena that cannot be accounted for within the bounds of such theories. Consequently social psychology still consists of dozens of self-contained but largely unrelated mini-theories (cf. Vallacher and Nowak 1994). Maybe the fragmented nature of contemporary social psychology simply reflects the complexity of human experience. Human experience of the relationship with a loved one, for example, is very different from that associated with strategic planning on a work project. The domain of social psychology covers phenomena at many different levels, from physiological activity up to cultural change; each level of analysis may require a different theoretical approach. The intuition that one needs diverse approaches to adequately capture processes occurring at different levels is common among both lay people and psychologists. It is conceivable that these various approaches actually cannot be meaningfully related to each other, in which case this book is doomed to meet the same fate as earlier over-arching theories.

Complexity science, however, suggests a different perspective on the relationship between phenomena at different levels. From this perspective, almost any phenomena may be understood as a system of interacting lower level elements (cf. Haken 1978; Holland 1995; Johnson 2001; Schuster 1984; Strogatz 2003; Weisbuch 1992). In analyzing a phenomenon, the task is to identify the rules of interaction among elements and to investigate how these rules promote the emergence of macro level phenomena. In fields as distinct as financial markets and cell metabolism, the perspective of complex systems science has revealed that apparently highly complex behavior at the macro level can emerge from very simple rules of influence between the system's elements. Moreover, such rules have been found to be remarkably similar, even for highly distinct phenomena.

In social psychology this perspective is less advanced than in some other social sciences, such as political sciences or economics. However, it has already proven useful, providing a new set of research tools as well as theoretical models of several distinct phenomena. Social psychology is a very broad discipline, though, and many topics and issues have yet to be addressed explicitly from a complexity perspective.

The complexity perspective on social psychological phenomena is often referred to as dynamical systems in social psychology (Vallacher and Nowak 1994) or dynamical social psychology (Nowak and Vallacher 1998a). The explicit concern with dynamics stresses the difference between the classical approach to social psychology, which concentrates on analyzing static phenomena, and the complexity approach (Vallacher et al. 2002b). Furthermore, since dynamical principles characterize complex systems generally, the dynamical approach to investigating personal and interpersonal experience holds potential for relating social psychology with related scientific areas.

Coherence in Social Psychological Systems

Complexity science applied to modeling social psychological systems posits two mechanisms for self-organization: within a level and between levels. Self-organization of elements at a lower level occurs through the application of rules. In turn, this enables the emergence of coherent psychological or social structures at the next higher level of social reality.

Emergence can occur from any level. As a coherent state of the system is reached at one level, it is recognized as such and emerges as an element at the next level up. Elements at the higher level may in turn become coherent leading to the emergence of yet higher-order properties. In theory, this process can continue until the system as a whole is characterized by a single macro-level state that functions in accordance with a single process. In practice, progressive integration is halted at a level where coherence stalls (cf. Simon and Holyoak 2002; Thagard 1989; Nowak et al. 2000).

The degree of coherence of a set of elements at one level is determined by the consistency of the implications arising from the application of the rules applied to the elements at that level. For example, a set of thoughts relevant to social judgment is coherent if, collectively, they convey an unequivocal evaluation of some person or social object (as a contrary example, think of an effective but ruthless head of state); a set of actions is coherent if they can be executed and lead to the performance of an activity, which in turn is comprehensive at a higher-level (for example, a gymnast putting together the micro actions for a triple salto). However, this process of ever higher-order coherence may stall well beneath that of a consistent evaluation or action mastery. Then social judgments will reflect a differentiated view (cf. Kunda and Thagard 1996) and activity will be characterized by the prepotency of uncoordinated acts (Vallacher and Wegner 1985, 1989).

Emotion and Affective Experience

A system may fail to achieve coherence for several reasons, such as the novelty or ambiguity of the inputs. There has then to be both some mechanism for the system to recognize failure to achieve coherence and also to proceed to solve the impasse.

Coherence in different socio-psychological systems has been modeled using attractor networks (Vallacher and Nowak 1999; Nowak and Vallacher 1998b). Attractor neural networks (Hopfield 1982) are connectionist networks in which there is continuous feedback between elements of the network. If a stimulus pattern is experienced as being in the basin of an attractor, i.e. similar to a known stimulus pattern, the signals arriving at most of the neurons from other neurons are consistent with the state of that neuron, so it does not have to send out a different signal. The state of other neurons is then drawn to the state specified by the attractor, i.e. the system is attracted to the attractor state; the energy of the network is at a minimum. However, when the network is far from an attractor the system is characterized by incoherence – the signals arriving at a given neuron from other neurons stimulate a change of state of that neuron which then changes its output thus potentially leading to the inputs to other neurons to change, causing them to change their state and so on (Lewenstein and Nowak 1989a, b). If this process continues there is a lack of coherence.

The number of elements changing state, e.g. neurons firing, is the 'coherence' of a network, so coherence provides a self-control feedback loop. Attractors taking a system to a coherent state reduce the amount of neuron firing or 'noise'. With familiar inputs which are comparatively coherent, the feedback loop results in lower noise levels (more coherence) until the stimulus is captured by an attractor. However, if the level of coherence falls, more elements are changing their state, which leads to a further decrease in the coherence of the network, in a positive feedback loop. This has the effect of progressively making even stronger attractors unable to capture the system, until at some point no attractor can capture the system's dynamics. This feedback loop enables the network to monitor its own progress: when there is little change or noise the network is coherent and the pattern is familiar, but when there are high levels of noise the network is incoherent and the pattern is unfamiliar.

But how do attractors arise? One method, reconsolidation, is first to seed the system with random attractors. When no attractor can capture the inputs to the network, the nearest attractor is moved towards the input or the current state, and the network restarted, repeatedly, until an attractor succeeds in capturing the network. Alternatively, Hebbian learning may be used. In Hebbian learning, the occurrence of the firing of one neuron followed by the firing of another to which it is connected leads to the reinforcement of the connection (the synapse) between the two neurons. Using this mechanism attractor networks can build new attractors, e.g. learn that a certain input pattern is familiar. From then on the occurrence of similar patterns are recognized by the system's new attractor with little delay.

Psychological processes have, however, not just one system but several, the main ones being perception, cognition, action and emotion. In general, within each of these systems and, where appropriate between these systems, there is a drive for coherence (e.g., Kruglanski and Webster 1996). Most accounts suggest that emotion serves to signal the state of coherence of the other social psychological systems but also the quality of coordination between them (e.g., Carver and Scheier 1999; Csikzentmihalyi 1990; Festinger 1957; Heider 1958; Higgins 2000; Kruglanski and Webster 1996; Mandler 1975; Simon 1967; Thagard and Nerb 2002; Vallacher and Nowak 1999; Vallacher et al. 1989; Winkielman and Cacioppo 2001). Not only do ambiguity of perception, uncertainty in judgment and uncoordinated actions, all give rise to negative emotional states, but so do discrepancies between perception and expectation, between mental representation and intended behavior and between intention and action. These negative emotions may be diffuse (e.g., arousal, agitation) or quite specific (e.g., disliking, guilt, self-consciousness).

Attractor networks have been used to model coherence in different sociopsychological systems (Vallacher and Nowak 1999). The units in an attractor network may be neurons, as in the example above, but can be mental identities, from cognitive elements (in the case of judgment) to individuals (in the case of group dynamics). Noise is interpreted as representing a variety of emotional states, including diffuse arousal, acute anxiety, self-conscious emotions such as embarrassment or guilt, or negative affect toward a stimulus, depending on the phenomenon. A negative emotion engages a second self-regulation feedback loop to restore coherence. For example, in the presence of a negative emotion linked specifically to an element of the system, the element may be disassembled into its sub-elements and attention transferred to a lower level to create a new element more appropriate to the current circumstance.

Consider actions, for example. A motor system may have been rendered ineffective by external influences by, for example the cues instigating action becoming ambiguous or conflicting, or by novel circumstances that disrupt the normal course of action. Then the action elements of the system can no longer be coordinated into an effective activity and therefore are no longer coherent. This instigates noise in the system which in turn is signaled by a negative emotion (e.g., anxiety, self-doubt). The effect of the negative emotion is to disassemble one of the actions into its lower-level acts. These acts can then be reassembled in a different configuration into a new action that may restore coherence to the action system. As a simple example, failure to undo the lid of a jar by hand will lead to finding other means of opening the jar, some of which may be in the person's repertory but, failing these, a new method may have to be invented, such as unusually asking someone's help.

System coherence and effective self-regulation, on the other hand, lead to positive emotion. When an action system, for example, functions autonomously it may produce an affective state commonly referred to as 'flow' (Csikzentmihalyi 1990). Positive emotions free up consciousness from monitoring the functioning of the system, and thus allow attention to move elsewhere, perhaps upward to assemble higher-level systems.

Positive and negative emotions are asymmetrical (cf. Cacioppo et al. 1997; Carver and Scheier 1999; Higgins 2000): whereas negative emotions focus attention on the internal workings of an incoherent system, positive emotions are associated with progression of consciousness to higher levels of integration. Effective self-regulation at a basic level (for example of an activity such as hitting a tennis ball) not only feels good, it also permits attention to the activity's meanings, such as its purpose, consequences, and implications (Vallacher and Wegner 1987).

The assembly of coherent lower-order systems into an element of a higher-order system and the repair of disrupted systems impart a dynamic quality to both affective and cognitive experience. Each time attention is diverted from a higher-order system to its lower-level elements, there is a press to reassemble the elements into a new higher-order pattern. As the resultant pattern may differ considerably from the previously disassembled pattern, there may be a qualitatively change of behavior. With each enactment of the disruption-repair cycle, then, there is potential for the creation of a new higher-order coherent system of self-regulation. From a dynamical perspective, the content of mind is open-ended and ever-changing, being a constructive process that fosters adaptation to changing demands and conditions.

The principles of self-organization leading to coherence and progressive integration may have a wide range of application, linking previously unrelated description of social phenomena. To date, however, research designed to illuminate the building of psychological structures in this fashion is limited to a few domains. We will now look at how these dynamics work in two of these: social influence and the self-concept.

Dynamic Model of Social Influence

The essences of the processes that enable individuals to coordinate their opinions, moods, evaluations, and behaviors in dyads or social groups can be captured by dynamic social impact theory (Nowak and Vallacher 1998a).

Every individual is influenced by their social environment. People experience a "great variety of changes in physiological states and subjective feelings, motives and emotions, cognitions and beliefs, values and behavior (...) as a result of the real, implied, or imagined presence or actions of other individuals" (cf. Latané 1981, p. 343). The influence that social environment has on an individual is called social impact.

Social Impact

The core of the dynamic social impact theory is based on an earlier static framework provided by Latané (1981). According to the static version of social impact theory mutual influences between individuals are moderated by three major factors: the strength of the source of influence, the distance between source and target, and the number of people involved, including both the influencing agents and the targets of the influence.

The strength of the source of influence is related to the salience, power, importance, or intensity of the source. It is determined by characteristics of the influencing agent such as their socio-economic features, the power relationship with the target, and transient characteristics specific for a given context, e.g. their motivation to persuade. For example, a person experiences a greater amount of impact, i.e. perceives a message as more persuasive, when the message, comes from a source that has high as opposed to low status (e.g. professor vs. student). The amount of impact experienced by the target person is proportional to the strength of the source of the influence.

The second determinant is the proximity in space or time between the target and the source of influence. It might be described as distance in the social space, reflecting the ease of communication and is called immediacy (Latané et al. 1995). Greater impact is experienced when immediacy is high. For example, news about events that happen in a close physical proximity to an individual attract more of their attention than news concerning distal events (Latané 1981). According to the social impact theory "spending time with, paying attention to, remembering, and being persuaded by someone should all decline with distance" (cf. Latané et al. 1995, p. 797).

The last factor in the model, the number of people involved, accounts for the effect both of multiple sources of influence affecting one and the same target and also a group of people being influenced. An increase in the number of agents influencing the target translates to an increase in the amount of impact experienced by the target, however not in a directly proportional manner. Based on empirical data and a review of previously published research findings Latané (1981) presented the following formula $I = s N^t$, $t < 1$, where the magnitude of impact, I, is equivalent to a power function of the number, N, of sources. The s accounts for specific factors constant in a given situation. The exponent t is usually approximately 0.5 as the amount of impact of a group of agents grows as a square root of the number of those agents (Latané 1981; Nowak and Vallacher 1998a). This means that the impact added by an additional influencing agent decreases as the number of influencing agents increases. It makes a large difference whether there is one or two influencing agents, but it does not matter that much anymore whether there are 33 or 34 of them.

When a group of people become a target of influence the actual amount of impact experienced by an individual in that group is a fraction of the influence exerted by the source. Research on diffusion of responsibility (e.g. Darley and Latané 1968) or social loafing (e.g. Latané et al. 1979) confirms that "social influence is divided among the individuals in a group" (Nowak and Vallacher 1998a, p. 225) which explains why a person in a group is often less likely to help or less willing to put effort into a collective action.

The social impact theory can be expressed by a mathematical function: social impact $= f(S I N)$, in which the amount of impact is a multiplicative function of the source strength (S), its immediacy (I), and the number of people involved (N).

Generally speaking, social impact theory (Latané 1981) provides a model useful to understand how people affect each other irrespective of the unique context of their interactions. Whether it is petition signing, stage fright, interest in news events, or bystander intervention, the social impact theory is able to extract and identify the meaningful forces that drive the emotions, cognitions, or behavior of an individual responding to the pressures of their social environment.

Dynamic Social Impact Theory

Social impact theory has been extended into a dynamic social impact theory (Nowak et al. 1990). The dynamic approach puts a new spin on the general principles of social impact and extends them to the dynamic aspects of social influence involving reciprocal interactions between the target of influences and their social environment. Nowak et al. (1990), applying the basic laws of social impact, show what happens in a society of spatially distributed individuals that are not only influenced by, but simultaneously exert an influence on, other members of population using a cellular automata model of opinion dynamics.

In this model social space is modeled as a two-dimensional grid with each cell representing an individual – see Fig. 1.1. Each individual is characterized by their opinion on an issue, either in binary, denoting whether they are 'for' or 'against' a particular position (Nowak et al. 1990), or as a continues variable

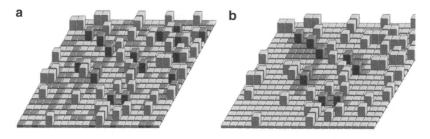

Fig. 1.1 (**a**) Initial distribution of opinions in the simulated group; (**b**) final equilibrium of opinions in the simulated group (Nowak et al. 2002)

(Nowak et al. 1993). The person's opinion is shown by the color of a given cell, and the strength of this opinion in a third dimension, by the height of a bar in a given cell. The higher the bar, the greater the persuasive strength of an individual. Location on a grid with respect to other members of the population reflects individual's position in the social space. Thus, distance between two individuals corresponds to immediacy.

The dynamics assumes that each individual tries to adopt an opinion that prevails at a given moment. They do this during social interactions (e.g. conversations) with other members of their social environment, by assessing how much support each opinion has. In a simulation at each step one individual is select for updating. Following the principle of immediacy the opinions of those who are closest to them are given most consideration. This includes the individual's own opinion, which is weighted most heavily. Opinions of strong individuals have a greater impact, reflecting the strength factor in the social impact principles. If an individual's currently held opinion is outweighed by the amount of social impact towards a different opinion, their opinion changes to match the prevailing one. This process is repeated for a next individual, usually chosen at random, and continues until there are no further changes in opinions.

Figure 1.1 presents representative results of the computer simulations. In Fig. 1.1a, there is a majority of 60 % (light gray) and a minority of 40 % (dark gray). The majority and minority members are initially randomly distributed, and each region has the same relative proportions of strong and weak members (high vs. low boxes). Figure 1.1b shows the equilibrium reached after six rounds of simulated discussion. Now the majority is 90 % and the minority is 10 %. The minority opinion has survived by forming clusters of like-minded people around strong minority individuals.

These two group-level outcomes – polarization and clustering – are commonly observed in computer simulations (cf. Latané et al. 1994) and are reminiscent of well-documented social processes. As a result of group discussion the average attitude in a group becomes polarized in the direction of the prevailing attitude. Polarization in the simulations reflects the greater influence of the majority opinion. In the initial random configuration (Fig. 1.1a), the average proportion of neighbors holding a given opinion corresponds to the proportion of this opinion in the total group. This means that the average group member is surrounded by more majority than minority members, which results in more minority members being converted to the majority position than vice versa. Some majority members are converted to the minority position, however, because they happen to be located close to an especially influential minority member, or because more minority members happen to be in this region.

The computer simulations reveal that rules at the individual-level underly the emergence of global order. Certain complex macro-level processes, such as group polarization or group coherence take place, not because of any macro-level process but as an effect of what is happening in relations on an individual-level. The cellular automata model developed shows that "simple laws about individual social reactions can, when applied reciprocally and recursively, predict emergent group effects" (cf. Nowak et al. 1990, p. 354). "No special process of greater majority persuasion is required to explain group polarization, nor any notion of greater minority influence to account for the fact that polarization is incomplete" (Nowak et al. 1990, p. 374). Local interactions lead to self-organization of the system, which results in the emergence of macro-level properties that are nowhere to be found in the lower-level of the system.

Clustering and Polarization

Clustering is pervasive in social life. Attitudes have been shown to cluster in residential neighborhoods (Festinger et al. 1950), for example, and pronounced clustering has been observed for political beliefs, religions, clothing fashions, and farming techniques. Clustering reflects the relatively strong influence exerted by an individual's neighbors. When opinions are distributed randomly, the sampling of opinions through social interaction provides a reasonably accurate portrait of the distribution of opinions in the larger society. When opinions are clustered, however, the same sampling process will yield a highly biased result. Because the opinions of those in the nearby vicinity are weighted the most heavily, the prevalence of one's own opinion is likely to be over-estimated. Hence, opinions that are in the minority in global terms can form a local majority. Individuals who hold a minority opinion are therefore likely to maintain this opinion in the belief that it represents a majority position.

Clustering occurs despite the press for coherence that is responsible for progressive integration in psychological systems. Three factors have been identified that effectively stall the integration process, preventing complete unification and hence preserving minority opinions in groups (Latané and Nowak 1997; Lewenstein et al. 1993; Nowak et al. 1996). Individual differences, first of all, are indispensable to the survival of minority clusters. By counteracting the sheer number of majority opinions, strong leaders stop minority clusters from decaying. As a result of social influence, moreover, individual differences in strength tend to become correlated with opinions. This is because the weakest minority members will most likely adopt the majority position, so that the average strength of the remaining minority members will grow over time at the expense of the majority. This scenario provides an explanation for why individuals advocating minority positions are often more influential than those advocating majority positions (cf. Moscovici et al. 1969).

The second factor is nonlinearity in attitude change. Abelson (1979) showed that when individuals move incrementally toward the opinions of their interaction partners, groups invariably become unified in their support of the majority opinion. In the model of dynamic social impact, however, attitudes change nonlinearly in accordance with a threshold function. Thus, individuals hold their opinion until social influence reaches a critical level, at which point they switch from one categorical position (e.g., pro) to the other (con). So whereas a linear change rule, which implies a normal distribution of opinions, promotes unification of opinions, a nonlinear change rule, which implies a bimodal distribution, can prevent complete unification and enable minority opinion to survive in clusters. Latané and Nowak (1994) have shown that a normal distribution tends to develop for relatively unimportant attitudes, but that a bimodal distribution is more often observed for attitudes of high personal importance. This suggests that consensus in a group can be achieved by decreasing the subjective importance of the topic.

The third factor is the geometry of the space in which individuals interact (Nowak et al. 1994). People do not communicate equally with all members of a group, nor are their interactions random. In the cellular automata model, different communication patterns can be approximated with different geometries of social space. In the limiting case, geometry is lacking altogether and interactions occur randomly between people. Under these conditions, minority opinion decays rapidly and the group converges on the majority position. Other geometries have been used to capture different communication patterns, and these have been shown to have predictable consequences for the fate of minority opinions. In real social settings, of course, several different geometries are likely to coexist and determine the emergence of opinion structure in groups. Even in a small town, the ready availability of telephones, e-mail, shopping malls, and common areas for recreation add many dimensions to the effective geometry in which interactions take place. The combined features of these geometries are certain to play significant roles in shaping the distribution of public opinion.

Self-Reflection and the Emergence of Self-Concept

Processes of self-organization and emergence are not unique to opinion formation. Another interesting example is a model of the dynamics of the self-concept called Society of Self (Nowak et al. 2000). The name of the model is meaningful and points to the parallelisms between rules underlying the organization and functioning of a society comprised of individuals and a self system comprised of knowledge elements. This gives us an inspiring analogy between two seemingly very distinct complex dynamical systems.

The self is the largest and most accessible structure in a person's mental system. Moreover, every facet of personal experience is potentially relevant to a person's self-understanding, from the details of their physical appearance to their self-perceived traits, values, and aspirations. The information relevant to the self that is encountered on a daily basis is vast and highly diverse both in content and valence, ranging from incidental events, feedback from significant others to success versus failure in personal pursuits. The nonstop exposure to distinct pieces of self-relevant information, often inconsistent, would seem to mitigate against the formation of a stable and coherent sense of self. Yet most people do develop relatively assured and coherent conceptions of themselves, especially concerning higher-order constructs such as traits, skills, and goals.

In addressing this anomaly, Nowak et al. (2000) conceptualized the self-structure as a system composed of cognitive elements, each representing a piece of information relevant to the self-concept, together with mechanisms of self-organization promoting coherence and stability in much the same way that mechanisms promote social consensus as just described in the preceding section. Although the elements of self-structure are diverse, they all carry a valence that can vary. Neighboring (thematically related) elements influence each other to adopt a common value. An element whose valence is incongruent with its neighboring elements may change its valence or the valence of its neighbors may change, thereby establishing an equivalent valence over the related elements. For example, the belief that one is distractible may take on positive rather than negative valence in the context of other self-perceived qualities that together convey an image of oneself as a creative scientist. This press for coherence generates subsets of self-relevant elements carrying the same or closely related valence. Furthermore, the self-structure becomes differentiated, with different regions stabilizing on different values of self-evaluation (e.g., Showers 1992). A person may have a coherent and positive view of themselves as a scholar, for example, and an equally coherent but negative view of themselves as an athlete.

The Society of Self Model

In the Society of Self cellular automata models this process of evaluative differentiation in a self-system. See Fig. 1.2a, b, in which each tile on a two-dimensional grid represents an individual element of self-awareness (e.g. memories relevant to a person's self-image). In addition to carrying an evaluative signature (positive valence is denoted by light colored tiles on the grid; negative valence is denoted by dark color) each element is also characterized by its relative importance for the self, which in the figures is denoted by the height of the tiles. More important aspects, those central to a person's self-concept, have a greater chance of influencing the evaluation of their neighboring elements, and are also more resistant to changing their own evaluative state under the influence of their neighbors.

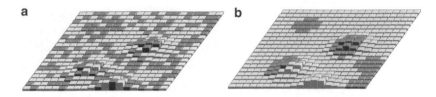

Fig. 1.2 Computer simulation of self-organization process. Each tile represents individual pieces of self-knowledge. *Light colored* tiles have positive evaluation, *dark colored* tiles have negative evaluation. The height of the tiles proportional to the relative importance of each element. The distance between the elements is shows how individual elements are related. (**a**) Shows undifferentiated system. (**b**) Shows differentiated system (Nowak et al. 2002)

With a system of elements, each having a binary evaluation, initially configured randomly (see Fig. 1.2a), simulation results showed that after the elements have adjusted their evaluative state to the valance then prevailing among the surrounding elements, the entire system is transformed into one composed of a number of larger areas with shared valence (see Fig. 1.2b).

The conglomeration of states with the same valence results from rules that dictate the behavior of individual elements rather than to any system level regulation (Vallacher and Nowak 2000). What these local interactions eventually produce depends both on the initial structure of the system (how the constituent parts are organized) and on the nature of influences between the elements (i.e. how individual elements change their evaluative state in order to achieve congruence with their neighbors). The simulation demonstrates the creation of evaluatively coherent substructures resulting in differentiation of the self-concept (see Fig. 1.2b). Differentiation of the self-concept emerges through self-organization of pieces of self-knowledge triggered by the press for evaluative integration.

The consequences of evaluative organization in the self-system were investigated in computer simulations which demonstrated that evaluative coherence has important consequences for two basic aspects of the self-system's functioning, namely, its dynamics and its sensitivity to external influences (Nowak and Vallacher 1998a).

The dynamism is directly related to the evaluative organization of the elements it operates upon, so that evaluative congruence between the elements is related to stability, and lack of evaluative coherence results in high dynamism. These findings were later corroborated by empirical experiments concerning self-evaluation dynamics (Vallacher et al. 2002a).

With respect to the sensitivity to external influences, simulation results have shown that a system comprised of elements that are not coherent is highly responsive to external influences; incoming information, for example, is then likely to change the evaluation of the elements. Whereas in an evaluatively differentiated self-concept, where groups of elements form clusters with similar valence, incoming information has less influence on the valence of elements, since the mutual influences between the elements make them more resistant to change (Nowak and Vallacher 1998a).

Simulations of the Society of Self Model

In the course of simulation, an element is chosen at random, and its valence is adjusted to be similar to its neighboring elements according to how much influence it receives from them. The sum of the valences of the neighboring elements, weighted by their importance, is computed and compared to the current valence of the element itself. If they are the same, the element's valence does not change. If they differ, the average importance for the self-concept of the neighboring elements is compared to the importance of the element under consideration; if the average is greater than the importance of this element, the valence is changed. So the valence of a relatively unimportant element is relatively likely to be changed, but it is less likely for the valence of a more important element of the self-concept to be changed.

This process is repeated for another randomly chosen element, and then again for another element, and so on, until the valence of each element has been considered. In the next simulation step, this process is repeated. The simulation continues until the system reaches an asymptote, when either there are no further changes in the state of elements (i.e., static equilibrium) or a stable pattern of changes in the system (i.e., dynamic equilibrium) emerges.

As Fig. 1.2 illustrates, the process of mutual adjustment of the valences of the elements of self-relevant information promotes the emergence of clusters. Self-relevant information whose valence is randomly distributed at the outset (left picture) forms well-defined clusters composed of elements that share a common valence (right picture). The emergence of evaluatively coherent clusters is due to the local nature of influence among elements. The self-system also becomes more polarized in overall evaluation, with more negative elements switching to positive valence than vice versa. In a disordered system, the proportion of positive and negative elements in a region roughly corresponds to the proportion of positive and negative elements in the entire structure. Hence, any given element is likely to be surrounded by more positive than negative elements, as it was randomly assigned at the start of the simulation, and thus is likely to experience greater influence in the positive direction. Once the self-structure has become clustered, however, most elements are surrounded by elements of the same valence, so that only the elements on the border of a cluster are subjected to conflicting influences.

The emergence of locally coherent regions tends to stabilize the self-system. This is because each element in a coherent region supports the current state of the other elements in that region. If the valence of such an element changes, the joint influence of the surrounding elements tends to return it to its original value. In effect, coherent regions function as attractors, where the value of each element is anchored in the value of the other elements of the coherent region. In an incoherent region, in contrast, the current state of an element is supported by some elements but undermined by others. Hence, when the valence of an element is influenced, say from new exogenous information, some of its surrounding elements support the new influence, whereas others resist it. If the valence of an element is changed by an outside influence, there will be little tendency for it to return to its original value since some of its neighboring elements are likely to support the new value.

Computer simulations provided support for coherence as a basis for the attracting tendencies of clusters. Integrated self-structures, those elements have the same valence, could withstand external influence and were able to rebound to their original valence after being perturbed. Unintegrated self-structures yielded more readily to influence and displayed weaker tendencies toward restoration of their original valence. Figure 1.2 also shows that although the proportion of positive elements increased, the negative elements that manage to survive tended to be more important and hence resistant to being changed. This is consistent with research on the negativity effect in judgment (e.g., Cacioppo et al. 1997; Pratto and John 1991; Skowronski and Carlston 1989). Although positive information tends to be more prevalent than negative information in cognitive structures, the relatively few elements of negative information tend to be more important.

Dynamical Minimalism

The subject matter of social psychology is obviously very complex, and a good theory must be able to account for this complexity. In the traditional approach to theory construction, the complexity of human thought and behavior is reflected in the complexity of the model, with many variables and complex interactions among them providing the starting point for an explanation of a phenomenon. The approach of dynamical minimalism (Nowak 2004), in contrast, tries to construct models in such a way that the observed complexity emerges from the simplest possible assumptions rather than being inherent in the model itself. The focus of this approach is on identifying the minimal set of realistic principles and mechanisms capable of producing the phenomenon under investigation. This perspective often assumes simple, almost trivial, assumptions at the level of individual elements, yet tries to reproduce the complexity of the phenomenon at the system level. Because the resultant theories provide simple explanations that nonetheless capture the complexity of human thought and behavior, this approach aims to maximize parsimony in theory construction without trivializing the phenomenon in question.

As emphasized in this chapter, simple rules governing the interactions among individual elements can generate very complex properties at the system level. For complexity to emerge from simple rules, though, the effects of some rules must interact over time with the inputs to other rules. Thus, a simple theory of a complex phenomenon is necessarily dynamic in nature. Dynamical minimalism is the approach of choice, then, when the relations among elements are nonlinear and the phenomenon displays self-organization and emergence. It is less essential for phenomena governed by linear dependencies, in which the potential for emergence is minimal. Of course, one can develop dynamic theories that do not propose emergence. Such a theory may be valid if one's interest is the nature of the dynamics per se and the theory's assumptions are verified empirically.

However, many relationships in social psychology, such as threshold phenomena, inverted-U relations, and statistical interactions, reflect nonlinearity and thus have the potential for emergence if the variables are embedded in a larger system that evolves over time (cf. Vallacher and Nowak 1997).

The minimalist approach provides a new perspective on the relation between micro and macro levels of description. From the perspective of reductionism, the properties at higher levels of description can be directly reduced to properties of elements at lower levels. The regularities at one level, in other words, result from regularities at a lower level. The relation between poverty and crime on the social level, for example, may be reduced to the relation between frustration and aggression operating at the level of individuals. In dynamical models, in contrast, behavioral rules at one level may generate wholly different behavior at a higher level. In the Society of Self model (Nowak et al. 2000), for example, basic rules regarding the integration of basic elements of self-knowledge have many interesting but unanticipated consequences for higher-level self-representation. We saw above that while the model assumes simply that each basic element takes on the prevailing valence of related elements, repeated iterations of this basic rule generate several interesting consequences at the global level of self-representation, such as the differentiation of self-structure into local regions of contrasting valence, and global properties (e.g., self-esteem) that are relatively immune to external influences and that can rebound after being challenged.

Building scientific theories that capitalize on emergence leads to an apparent paradox. By definition, emergence refers to principles on the system level that cannot be derived by reasoning about the knowledge of the system's elements. In a system characterized by emergence, then, how can the knowledge of lower-level elements serve as an explanation of higher-level properties? The answer highlights the crucial role played by computer simulations in the dynamical approach. Computer models enable one to specify properties of elements and the rules of interaction among them. When the elements interact according to the specified rules, behaviors may be observed at the system level that were not assumed on the level of individual elements. Indeed, a primary reason for constructing computer simulations is to identify the emergent consequences of basic rules. Computer simulations thus allow for a theory formulated under one level of psychological reality to be tested at a different level of psychological reality.

Computer simulations are essential for another reason. Many specific properties of individual elements and their interactions have only minimal effects on properties at the system level. Indeed, the basic elements themselves are typically uninteresting, even trivial in nature. What matters are some very basic properties of the elements and the patterns of interaction among the elements. Consider, for example, the emergence of public opinion modeled with cellular automata described above. This model characterizes each individual in terms of only three properties: their location on a two-dimensional grid, their attitude (e.g., pro vs. con) on some topic, and their persuasive strength. Clearly, any individual is far more complex than this. Many nuances of individual variability (e.g., idiosyncratic traits), however, have little, if any, impact on the dynamics of public opinion formation in a social group. So although this model does not do justice to the complexity of individuals, it captures the essential features that are responsible for the emergence of public opinion.

The goal of dynamical minimalism is to build a model that incorporates just those variables that are crucial for the emergence of macro level properties. In computer simulation models, one can systematically vary the assumptions concerning different properties of elements and their interactions, and observe which assumptions result in important changes at the macro level. The characteristics that do not have consequences at the system level can be disregarded and omitted from the model. In effect, computer simulations function as a sieve that distills the minimal set of components and their interactions that constitute the essence of the phenomenon of interest.

While computer simulations play a pivotal role in dynamical minimalism, the theory's assumptions must be empirically verified. Computer simulations are, moreover, useful in identifying the crucial assumptions of a theoretical model and thus provide a direction for empirical efforts. In a common scenario, computer simulations of processes assumed to operate at a lower level may be used to investigate the consequences of these processes at a higher level. These consequences, in turn, may function as hypotheses to be tested in empirical research. Dynamic social impact theory (Nowak et al. 1990), for example, was used to derive predictions concerning spatial-temporal patterns of social change processes. The existence of these hypothesized patterns was then assessed in statistical data concerning patterns of entrepreneurship and voting patterns in Poland following the fall of Communism in the 1990s (Nowak et al. 2005). In addition to testing a model's assumptions, empirical tests can also be used to refine the model. The refined model, in turn, can then be implemented in computer simulations, the results of which can provide further hypotheses to be tested empirically. This loop between theory, computer simulation, and empirical research is instrumental to the progress of scientific social psychology.

The Trajectory Ahead

The construction of theories in social psychology can itself be viewed in dynamical terms, such that individual researchers influence one another over time in an attempt to achieve consensus on the nature of human experience. But despite the progressive coherence that emerges over time by virtue of self-organization, complex systems rarely attain complete integration. Hence, it is unreasonable to expect that a discipline as diverse and multi-faceted as social psychology will reach a stable equilibrium, with a single set of immutable principles capturing all the nuances of personal and interpersonal function. Complex systems are inherently dynamic – they constantly evolve and undergo transformations by virtue of their intrinsic dynamics and in response to incoming information and outside influences. At this point in time, then, the promise of theoretical coherence is an optimistic extension of a current temporal trajectory. So, although we anticipate that dynamical principles and methods will emerge as the paradigmatic foundation for social psychology, we also anticipate that the field will undergo repeated episodes of disassembly and reconfiguration with respect to specific theories and research strategies in the years to come. Far from undermining the promise of dynamical social psychology, this trajectory provides ironic testament to the viability and generality of the dynamical perspective on human experience.

References

Abelson, R.P.: Social clusters and opinion clusters. In: Holland, P.W., Leinhardt, S. (eds.) Perspectives in Social Network Research, pp. 239–256. Academic, New York (1979)

Cacioppo, J.T., Gardner, W.L., Berntson, G.G.: Beyond bipolar conceptualizations and measures: the case of attitudes and evaluative space. Pers. Soc. Psychol. Rev. **1**, 3–25 (1997)

Carver, C.S., Scheier, M.F.: Themes and issues in the self-regulation of behavior. In: Wyer Jr., R. S. (ed.) Advances in Social Cognition, vol. 12, pp. 1–105. Erlbaum, Mahwah (1999)

Csikzentmihalyi, M.: Flow: The Psychology of Optimal Experience. Harper & Row, New York (1990)

Darley, J.M., Latané, B.: Bystrander intervention in emergencies: diffusion of responsibility. J. Pers. Soc. Psychol. **8**, 377–383 (1968)

Festinger, L.: A Theory of Cognitive Dissonance. Row, Peterson, Evanston (1957)

Festinger, L., Schachter, S., Back, K.: Social Pressures in Informal Groups. Stanford University Press, Stanford (1950)

Haken, H.: Synergetics. Springer, Berlin (1978)

Heider, F.: The Psychology of Interpersonal Relations. Wiley, New York (1958)

Higgins, E.T.: Making a good decision: value from "fit.". Am. Psychol. **55**, 1217–1230 (2000)

Holland, J.H.: Emergence: From Chaos to Order. Addison-Wesley, Reading (1995)

Hopfield, J.J.: Neural networks and physical systems with emergent collective computational abilities. Proc. Natl. Acad. Sci. **79**, 2554–2558 (1982)

Johnson, S.: Emergence: The Connected Lives of Ants, Brains, Cities, and Software. Scribner, New York (2001)

Kruglanski, A.W., Webster, D.M.: Motivated closing of the mind: "Seizing" and "freezing". Psychol. Rev. **103**, 263–283 (1996)

Kunda, Z., Thagard, P.: Forming impressions from stereotypes, traits, and behaviors: a parallel-constraint-satisfaction theory. Psychol. Rev. **103**, 284–308 (1996)

Latané, B.: The psychology of social impact. Am. Psychol. **36**, 343–356 (1981)

Latané, B., Nowak, A.: Attitudes as catastrophes: from dimensions to categories with increasing involvement. In: Vallacher, R.R., Nowak, A. (eds.) Dynamical Systems in Social Psychology, pp. 219–249. Academic, San Diego (1994)

Latané, B., Nowak, A.: The causes of polarization and clustering in social groups. Prog. Commun. Sci. **13**, 43–75 (1997)

Latané, B., Williams, K., Harkins, S.: Many hands make light the work: the causes and consequences of social loafing. J. Pers. Soc. Psychol. **37**(6), 822–832 (1979)

Latané, B., Nowak, A., Liu, J.: Measuring emergent social phenomena: dynamism, polarization and clustering as order parameters of social systems. Behav. Sci. **39**, 1–24 (1994)

Latané, B., Liu, J., Nowak, A., Bonavento, M., Zheng, L.: Distance matters: physical distance and social impact. Pers. Soc. Psychol. Bull. **21**, 795–805 (1995)

Lewenstein, M., Nowak, A.: Fully connected neural networks with self-control of noise levels. Phys. Rev. Lett. **62**, 225–229 (1989a)

Lewenstein, M., Nowak, A.: Recognition with self-control in neural networks. Phys. Rev. A **40**, 4652–4664 (1989b)

Lewenstein, M., Nowak, A., Latané, B.: Statistical mechanics of social impact. Phys. Rev. A **45**, 703–716 (1993)

Mandler, G.: Mind and Emotion. Wiley, New York (1975)

Moscovici, S., Lage, E., Naffrechoux, M.: Influence of a consistent minority on the responses of a majority in a color perception task. Sociometry **32**, 365–380 (1969)

Nowak, A.: Dynamical minimalism: Why less is more in psychology. Pers. Soc. Psychol. Rev. **8**, 183–192 (2004)

Nowak, A., Vallacher, R.R.: Dynamical Social Psychology. Guilford, New York (1998a)

Nowak, A., Vallacher, R.R.: Toward computational social psychology: cellular automata and neural network models of interpersonal dynamics. In: Read, S.J., Miller, L.C. (eds.) Connectionist Models of Social Reasoning and Social Behavior, pp. 277–311. Erlbaum, Mahwah (1998b)

Nowak, A., Szamrej, J., Latané, B.: From private attitude to public opinion: a dynamic theory of social impact. Psychol. Rev. **97**, 362–376 (1990)

Nowak, A., Lewenstein, M., Szamrej, J.: Bąble modelem przemian spolecznych (Bubbles – a model of social transition). Swiat Nauki (Scientific American Polish Edition) **12**, 16–25 (1993)

Nowak, A., Latané, B., Lewenstein, M.: Social dilemmas exist in space. In: Schulz, U., Albers, W., Mueller, U. (eds.) Social Dilemmas and Cooperation, pp. 114–131. Springer, Heidelberg (1994)

Nowak, A., Lewenstein, M., Frejlak, P.: Dynamics of public opinion and social change. In: Hegselman, R., Pietgen, H.O. (eds.) Modeling Social Dynamics: Order, Chaos, and Complexity, pp. 54–78. Helbin, Vienna (1996)

Nowak, A., Vallacher, R.R., Tesser, A., Borkowski, W.: Society of self: the emergence of collective properties in self-structure. Psychol. Rev. **107**, 39–61 (2000)

Nowak, A., Strawinska, U., Johnson, S., Vallacher, R.R.: Zaburzenia samoregulacji przy niskim poczuciu własnej wartości i w depresji (Malfunctions of self-regulation in low self-esteem and in depression.) Kolokwia Psychol (2005)

Pratto, F., John, O.P.: Automatic vigilance: the attention grabbing power of negative information. J. Pers. Soc. Psychol. **61**, 380–391 (1991)

Schuster, H.G.: Deterministic Chaos. Physik Verlag, Vienna (1984)

Showers, C.J.: Compartmentalization of positive and negative self-knowledge: keeping bad apples out of the bunch. J. Pers. Soc. Psychol. **62**, 1036–1049 (1992)

Simon, H.A.: Motivational and emotional controls of cognition. Psychol. Rev. **74**, 29–39 (1967)

Simon, D., Holyoak, K.J.: Structural dynamics of cognition: from consistency theories to constraint satisfaction. Pers. Soc. Psychol. Rev. **6**, 283–294 (2002)

Skowronski, J.J., Carlston, D.E.: Negativity and extremity biases in impression formation: a review of explanations. Psychol. Bull. **105**, 131–142 (1989)

Strogatz, S.: Sync: The Emerging Science of Spontaneous Order. Hyperion Books, New York (2003)

Thagard, P.: Explanatory coherence. Behav. Brain Sci. **12**, 435–467 (1989)

Thagard, P., Nerb, J.: Emotional gestalts: appraisal, change, and the dynamics of affect. Pers.Soc. Psychol. Rev. **6**, 274–282 (2002)

Vallacher, R.R., Nowak, A.: The chaos in social psychology. In: Vallacher, R.R., Nowak, A. (eds.) Dynamical Systems in Social Psychology, pp. 1–16. Academic, San Diego (1994)

Vallacher, R.R., Nowak, A.: The emergence of dynamical social psychology. Psychol. Inq. **4**, 73–99 (1997)

Vallacher, R.R., Nowak, A.: The dynamics of self-regulation. In: Wyer Jr., R.S. (ed.) Advances in Social Cognition, vol. 12, pp. 241–259. Lawrence Erlbaum, Mahwah (1999)

Vallacher, R.R., Nowak, A.: Landscapes of self-reflection: mapping the peaks and valleys of personal assessment. In: Tesser, A., Felson, R., Suls, J. (eds.) Psychological Perspectives on Self and Identity, pp. 35–65. American Psychological Association, Washington, DC (2000)

Vallacher, R.R., Wegner, D.M.: A Theory of Action Identification. Lawrence Erlbaum, Hillsdale (1985)

Vallacher, R.R., Wegner, D.M.: What do people think they're doing? Action identification and human behavior. Psychol. Rev. **94**, 3–15 (1987)

Vallacher, R.R., Wegner, D.M.: Levels of personal agency: individual variation in action identification. J. Pers. Soc. Psychol. **57**, 660–671 (1989)

Vallacher, R.R., Wegner, D.M., Somoza, M.P.: That's easy for you to say: action identification and speech fluency. J. Pers. Soc. Psychol. **56**, 199–208 (1989)

Vallacher, R.R., Nowak, A., Froehlich, M., Rockloff, M.: The dynamics of self-evaluation. Pers. Soc. Psychol. Rev. **6**, 370–379 (2002a)

Vallacher, R.R., Read, S.J., Nowak, A. (eds.): The dynamical perspective in social psychology. Pers. Soc. Psychol. Rev. **6**(special issue), 264–273 (2002b)

Weisbuch, G.: Complex Systems Dynamics. Addison-Wesley, Redwood City (1992)

Winkielman, P., Cacioppo, J.T.: Mind at ease puts a smile on the face: psychophysiological evidence that processing facilitation leads to positive affect. J. Pers. Soc. Psychol. **81**, 989–1000 (2001)

Chapter 2
Understanding Cognition Through Functional Connectivity

Agnieszka Rychwalska

Abstract With every word that you read on this page, your brain orchestrates a symphony of electrical sounds – millions of neurons perform at the same time and billions of synapses coordinate their sounds. If you make yourself a break and start preparing a coffee, a new array of neural musicians will become active. While we know right now quite well how these functions that you perform are segregated in the brain – that is, which set of neurons activates to enable your reading and which to make you remember where you put the coffee jar – it still remains a challenge to understand how the brain integrates separated tasks into a coherent function. How does it happen that the letters form a word in your mind and the words form a meaningful sentence? How do you coordinate the movement of your hands when you reach for the cup with one and for the coffee pot with the other? New tools made available by complexity sciences – the modern network theory – give us a unique chance to describe and measure the integration of information in the brain that is crucial for any function it performs.

It has been a century now since we learned that the brain is a network. In the beginning of the twentieth century Santiago Ramon y Cajal ended a long standing debate on whether the neural system is composed of separate cells or a continuous mass of tissue; with histological imaging y Cajal and Azoulay (1911) proved without doubt that it was the former. These cells communicate with each other at special junction points – the synapses. While a substantial amount of research has been devoted to tracking specific connections and pathways within this web of neurons, the network nature of the brain as such has only recently been subjected to investigation with network tools. Huge advances in brain imaging on the one hand and complex systems tools on the other finally enable us to quantitatively describe – with the precision of formal sciences – the structure and complexity of brain networks. This, in turn, might provide us with an unprecedented insight into the integrative function of the brain.

A. Rychwalska
Institute for Social Studies, University of Warsaw, Stawki 5/7, 00183 Warsaw, Poland
e-mail: izziaczek@yahoo.com

A. Nowak et al. (eds.), *Complex Human Dynamics*, Understanding Complex Systems,
DOI 10.1007/978-3-642-31436-0_2, © Springer-Verlag Berlin Heidelberg 2013

Information Integration in the Brain

The mechanisms of information integration in the brain are crucially important for our understanding of the biological basis of cognition and – more generally – the mind. Our phenomenological, subjective experience of cognitive function involves different levels of organization of processes. For example, the process of recognition starts with the combination of tiny perceptual inputs into meaningful shapes. These are then organized into bigger wholes – objects – that, in turn, may be mentally juxtaposed against stored memories about both episodic encounters with similar items as well as encoded semantic knowledge. A vital part in this process is the formation of functional units at different levels of description which, once assembled, can be operated upon without the need of analyzing their parts. That is, once a group of lines form a mental representation of a chair – a functional unit of a higher level – we do not need to operate any longer on single lines. Similarly, if the chair and all other furniture along with its placement become integrated into the mental dining room we can proceed to thinking about the family dinner without any need of disassembling the picture.

Clearly, mental processing depends to a large degree on organizing items into functional units. Even more importantly, our stream of consciousness undergoes a similar process. Single thoughts are organized into opinions, emotions, attitudes, memories, etc. which then become plans, schemas, narrations that – at the highest level – form the self. If the brain is the source of cognition, consciousness and the mind, there needs to exist neural mechanisms that provide for an analogous process of functional integration of information processing.

The anatomical structure of the brain is functionally segregated. That is, we can establish which regions – brain structures or areas – are selectively activated in the course of a specific function. In particular, the early perceptual regions are precisely mapped with regard to their function. For instance, in the primary visual cortex cortical columns are tuned to respond (by increasing the frequency of action potentials they produce) to very well defined stimuli in the visual field, e.g. gratings at specific angles (Hubel and Wiesel 1968). In areas that are further along the information processing pathway – secondary visual cortex or multimodal cortex – these preferences become more complex to combine information incoming from various other regions. Thus, the activation that spreads through the cortex from the external, perceptual input engages more and more functionally specific structures. Complex cognitive tasks activate a large set of brain regions to combine many functions to meet the demands of the task. While this segregation of function seems intricate enough to accommodate even composite cognitive processes, there still remains a question of how all this segregated information becomes integrated into coherent representation upon which we perform cognitive functions.

The most prominent – almost implicit – approach to information integration in the brain assumes that the mere act of transferring impulses from functionally specific areas to other, more complex ones, constitutes integration. That is, if a color specific region together with a shape specific region send impulses to a color and shape specific region, and manage to activate it (increase its rate of firing action potentials), it means that color and shape information has been integrated. There are several shortcomings to this line of reasoning, though. First, there is no single area in the brain where impulses from all other regions converge. In practical terms it means that there is no single steering centre in the brain; no area where all representations are formed and acted upon. Rather, the global, widespread activation levels seem to carry this function. Second, if the number of total possible combinations of input features were to be considered it is clear that there are not enough functionally specific areas to account for it. However complex the structure of functional segregation may be, alone it is not enough to explain how the brain performs its functions. The so called binding problem provides a simple example why activation spreading through functionally specialized regions does not suffice to reproduce behaviorally observable processes.

The Binding Problem

If you are presented with two colorful objects – let us say a green ball and a red cube – you do not have any problems in telling which object is of what color. Since there are color and shape specific areas in the visual cortex, we may say (even though it is not entirely realistic) that in such a situation four separate groups of neurons will be highly active – those responding to shapes of sphere and cube and those responding to the colors red and green. However, the same sets of neurons would be active in the reverse situation – when a red ball and a green cube would be present. How does the brain 'know' which color belongs to which object so that your mind image represents the reality? How are properties, such as shape and color, integrated into a whole without any mistake? Clearly, activation of functionally specific areas is not enough and another mechanism is in action.

Recent developments in neurophysiology show that a possible integrative mechanism lies in the precise timing of neural electrical activity (Singer 1999). According to this assumption synchronous firing of action potentials by two cells or cell ensembles means that they are activated by the same object. In other words, precise correlation of activity integrates the information that is processed by specialized groups of cells. This provides a solution to the binding problem – in the situation described above correlated activity of cells coding color green and a sphere means that there is a green ball in the visual field. Similarly, synchronous firing of cells responding to red and cube encodes a red cube. If a reverse combination was present in the visual field, correlations would appear between cells for green and cube as well as for red and sphere (see Fig. 2.1 for a graphical representation).

Neural Synchrony

Correlated neural activity has been found across many topological scales and in a variety of motor, perceptual and cognitive activities both in human and animal subjects. The synchronous activity can be either evoked by the incoming stimuli

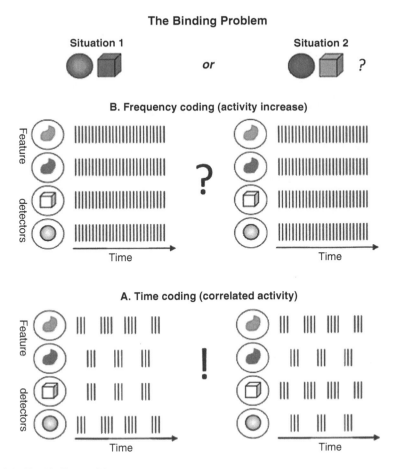

Fig. 2.1 The binding problem

(that is, time locked to the moment of presentation) or induced by it (not time locked, usually appearing after a time lag in the order of 100–300 ms). Correlation of signals within a group of cells is very often accompanied by oscillations at specific frequency ranges. It has been proposed that different frequencies of correlated activity can serve as an integrative function for different scales of processing.

It is worth noting that the time dependencies between neural signals referred to here as 'correlated activity' can be of various nature and therefore subject to different measurements. For example, it can be the precise timing of action potentials between cells or group of cells but also cross-coherence of aggregate electrical activity, such as local field potentials (LFP), post-synaptic potentials or electro-encephalograph signals.

The Temporal Correlation Hypothesis

An integrated approach to the role of correlated neural activity in information integration is presented in the temporal correlation hypothesis, the assumptions of which were first proposed by von der Malsburg as an answer to the problem of the combinatorial nature of feature integration (von der Malsburg 1994). A more comprehensive approach, along with the name, was formulated by Singer and Gray (1995). The main proposition is that in every neural system where information processing is distributed, the solution to the binding problems lies in the temporal characteristics of the neural activity. As opposed to previous attempts at explaining information integration in the brain, where the most important feature of neural activity was the frequency of firing (frequency coding), the temporal correlation hypothesis seeks a solution in the precise timing of action potentials (time coding). The main prediction that follows from these assumptions is that each group of neurons that process information about the same object (as is in the case of perception) or the same cognitive process (more generally), will share the same temporal pattern of sending action potentials – in other words, will synchronize their activity. This process of synchronization, if it is to be realistic, needs to be enacted to within a millisecond range.

A great advantage of this model lies in the fact that it enables integration of multiple objects, or functions, at the same time. This concurrence may be possible either by differentiating temporal patterns (including, inter alia, different frequencies) as well as phase differences. It is also possible for one cell or cell group to partake in more than one integrated unit at the same time – for example, by synchronizing with other cell assemblies at harmonic frequencies. As such, the problem of coding multiple representations (even those that share some features) by the same group of neurons at the same time is solved.

The temporal correlation hypothesis does not stand in opposition to frequency coding, however. Rather, it proposes a complementary mechanism of integration that enhances the coding capacity of neural systems. In addition, it explains why integrated wholes may be easily acted upon at higher levels of information processing: a synchronized assembly forms a functional unit at a higher level which is easily distinguishable from other cells. In terms of neural activity this ease of distinguishing is the ability of a synchronized signal to quickly activate the next levels of information processing. Thanks to the process of temporal summation of signals, a neuron is far more likely to produce an action potential if incoming signals from other cells are synchronized. This way, synchronized cells are more visible than unsynchronized ones, even if there are more of the latter.

At each level of information processing synchronized groups form functional units that integrate more and more complex structures. Such groups may be formed within sensory areas with a single modality but also across modalities; they may also join distant processing centers in the brain. Such far reaching synchronies have been proposed as the basis of multimodal perception, sensory awareness or momentary consciousness ('scene').

The micro level assumptions of the temporal correlation hypothesis have their origins in Hebb's propositions. A synchronized neural group constitutes a cell assembly as proposed by Hebb (1949), in which activation of a single cell, through intra-group connections, activates the whole cell assembly. This in turn strengthens the connections – Hebb's synapses – according to the rule that concurrent activation fosters stronger connections. Stronger connections, again, increase the chance for simultaneous activation, closing the feedback loop. In contrast to Hebb's postulates, the temporal correlation hypothesis does not rely on the formation of stable connections. Rather, it proposes temporal strengthening of synapses (LTP – long term potentiation and LTD – long term depression) as a mechanism to facilitate the realization of synchronized functional units. This way, they are dynamical formations, appearing for a short, distinct time and disassembling shortly thereafter to enable the creation of a new configuration of functional units.

Empirical Findings on Neural Synchrony

The most influential series of studies to confirm the predictions of temporal correlation hypothesis was designed by Singer, Gray and collaborators. In their first study they showed that a grating of a specific angle elicits synchronization of a cortical area that responds (i.e. increases the frequency of firing) preferably to this angle, while other areas remain unsynchronized (Gray et al. 1989). Further, the authors stimulated neighboring areas in the visual cortex (i.e. responding preferably to areas in the visual field that are next to each other) with moving gratings. If the gratings moved in different directions in each field, the cells responded with increased frequency of firing and intra-areal synchrony. If the stimuli moved in the same direction, a weak inter areal synchrony could be observed. However, if only a single grating was used that spanned both spatial fields, cells in both areas synchronized their activity. This result proves that synchrony is indeed related to cells processing information about the same object being 'tagged' as such. More-over, the level of synchrony is related to the extent to which stimuli form a coherent whole – if they can be viewed as a single object, synchronization is strongest, if they share some features (e.g. direction of movement), synchronization is weaker and finally, if they are totally separate, synchronization ceases.

A similar result was obtained in a follow up study, in which the authors recorded the activity of cells from the left and right hemispheres that shared similar receptive fields near the centre of the visual field (Engel et al. 1991). Again, a single object moving through the field elicited inter areal synchronization of activity, this time spanning both hemispheres. A section of the corpus callosum abolished this synchrony, leading the authors to the conclusion that the zero-lag synchrony was effectuated through inter hemispheric connections rather than being stimulated by a single subcortical structure. Moreover, this result proves that synchrony is not affected by the length of connections between the areas in question, yielding similar, millisecond precision both within a single cell column as well as across hemispheres.

An important result was found in a study by Fries et al. (2002). Strabismic cats (that had undergone a section of corpus callosum at an early age) were shown distinct stimuli to each eye (binocular rivalry). In such a situation, only one stimulus can be selected for processing at a time. The inter-ocular rivalry was controlled by increasing the saliency of one stimulus over the other, e.g. by setting a specific time offset or increasing the contrast. The actual selection of one stimulus was further controlled by comparing eye movements. The stimuli used were gratings of different angles that elicit responses from a range of visual cortical areas whose multi-unit activity (MUA – action potential trains from many cells) was recorded. Whenever the rivalry resolved in favor of the stimulus preferred by specific areas, their activity was synchronous – in contrast to the areas preferring the unselected stimulus. However, the firing rate – frequency of action potentials – was independent of synchrony. That is, since both stimuli were present at all times, they elicited increased activity in all cells responding to those stimuli. What distinguished – on the level of neural activity – the stimulus chosen for awareness was the level of synchrony of the cells responding to it. In fact, the frequency rate was biased towards the unselected stimulus, leading the authors to conclude that synchrony could be translated into increased frequency at some later stages of processing. The results of this study prove that time and frequency coding are separate mechanisms of information integration that in fact complement each other, with the time coding being crucial for awareness processes.

An amassed amount of experimental data demonstrate that synchronous oscillations are present in cortical activity across a variety of perceptual and motor tasks: in visual processing in monkey (Eckhorn 1994; Frien et al. 1994), in auditory cortex of humans during the presentation of stimuli to the dominant ear (Pantev and Elbert 1994), during auditory tasks in cats (Csépe et al. 1994); in sensori-motor cortex of both human and monkey during muscle movement and reaching for objects (Murthy et al. 1994; Murthy and Fetz 1992), in human sensory cortex while discriminating the stimulated finger (Desmedt and Tomberg 1994), and in visual cortex in humans during a visual field search (Tallon-Baudry et al. 1997).

Of exceptional significance are studies which point to the crucial role of neural synchrony in cognitive tasks. Von Stein et al. (1999) have shown that independent of the modality of the presented stimulus (visual – image, auditory – word, visual – word) long range correlations of activity appear in the temporal and parietal cortices. The authors conclude that this coordinated activity is crucial for multi-modal processing of incoming information. Von Stein and Sarnthein (2000) have further proposed that long range correlations – used for more general integration – typically involve lower frequency bands, while short range synchrony within specific areas and devoted to lower level integration, such as in perception, rely on higher frequencies (specifically, the gamma band). Similarly, Wrobel et al. (2007) proposed that while perceptual processes relate to gamma band synchrony, attention requires synchrony in the beta frequency band.

Pulvermüller et al. (1994) have shown that the analysis of words elicits stronger synchronous oscillations in the left hemisphere (all subjects were right-handed) than nonsense letter combinations. Miltner et al. (1999) discovered increased coordination of activity between visual and sensori-motor cortices in an anticipa-tory learning task, where a visual cue warned before an electric shock. Fell et al. (2001) have shown that successful, as compared to unsuccessful, learning involves a specific pattern of coordination of activity between structures crucial for memory formation – an initial increase of correlation that is followed by a decrease as the task nears conclusion.

A very important study by Rodriguez et al. (1999) has shown that long range correlations are present in cognitive integration of information. Subjects were presented with 'moony' faces – simplified, black and white images of human faces that are easily recognized when presented upright and unrecognizable when shown upside down. Such stimuli are a perfect example of a Gestalt – an indivisible figure that is always perceived as a whole. Gestalt psychology describes a set of properties of such figures, e.g. top-down influence of the whole on its parts and bottom-up mechanisms of integration of parts into a whole (Kohler 1929). Gestalt figures are, therefore, specifically appropriate for studying information integration during cogni-tion. In the case of recognizing faces, as opposed to being unable to recognize any shape, a specific pattern of correlations was observed in the EEG signal. First, many long-range correlations appeared between the recording sites, specifically between the parieto-occipital area and the frontal and temporal regions. Then, the synchrony decreased to be finally followed by another increase, this time mostly in the parietal areas. The latter increase in coordination preceded a motor response – for which parietal activity is crucial – by which the subjects indicated whether they had recognized a face in the image. The results of this study show that integration of information into a coherent whole depends on a spatio-temporal pattern of neural synchrony. Moreover, the authors point to the fact that a period of desynchronization between distinct cognitive tasks (which was not accompanied by any decline in activity levels) might be crucial for switching from one function to another.

Of vital importance for our understanding of the role of coordinated neural activity in cognitive function are the results of a study by Tononi et al. (1998) on the coordination patterns during conscious perception. Subjects were presented with blinking gratings in a binocular rivalry condition. The stimuli for each eye had different frequencies of flickering. In such a setting, conscious perception is possible for only one stimulus at a time and naturally switches from one eye to the other every three seconds. A flickering stimulus elicits electrical neural activity of the same frequency, therefore it was possible to assess the awareness of a particular stimulus by measuring the strength of the frequency of each stimulus in the MEG (magneto-encephalograph) signals. The results show that conscious perception elicits a far more distributed activity (the activity of a given frequency spans many areas) than a non-consciously perceived image. What is more important, this distributed activity is accompanied by many long range correlations appearing between recording sites. This result shows again that information integration by synchrony is vital for conscious cognition.

One crucial study offers an experimental verification – as opposed to mere correlation – of the role of neural synchronization in perception. Stopfer et al. (1997) impaired neural ability to synchronize in bees by induction of picrotoxin, which is known to suppress synchronization. The bees were then presented with various odors that normally elicit 30 Hz synchronized oscillations in the olfactory bulb. Bees that were treated with picrotoxin were unable to distinguish similar odors while their ability to differentiate between quite distinct odors was left unimpaired. This study proves that synchronization plays a causal role in at least some perceptual tasks.

Functional Connectivity

The temporal correlation hypothesis – that summarizes the role of synchrony in neural processing – states that synchronized activity is tantamount to integration of information processed by separate, functionally specific units. However, the process of integrating incoming information is not a sequence of synchronization events between pairs of brain loci. Rather, any cognitive action requires a vast amount of information to be integrated simultaneously. This means that at the same moment of time many areas synchronize their activity, either in pairs or in larger groups. What we observe then, as witnessed in many of the studies presented above, is an array of areas connected by correlations between their electrical signals. Such connections are termed 'functional connections' – as opposed to anatomical ones – and form functional connectivity.

A further distinction is between 'functional' and 'effective' connectivity (Aertsen et al. 1989). The first describes time dependencies observable in the electrical activity of given sources. The latter term is used to underscore the causal relation between two signals – if one region elicits correlated activity in the other, an effective link forms between them. In case of functional links, this is not necessarily the case – the correlation of signals between two areas may be the effect of a third region imposing its activity pattern on both. Functional connectivity is thus a wider notion that encompasses effective connectivity and, so called, spurious correlations. However, due to technical limitations it is very often impossible to establish direct causal relations between measured signals and therefore it is functional rather than effective connectivity that is subject to investigation in most studies.

To study the role of functional connectivity in information integration and more general cognitive functions, it is necessary to extract functional connections from brain imaging data. To this end a cross-coherence matrix of interdependencies between recording sites is extracted from raw signals (see also Stam and Reijneveld 2007). This can be either coherence between haemo-dynamic responses in the case of fMRI (functional magnetic resonance imaging) recording or between potentials recorded through MEG, EEG or LFP. A threshold value is then applied to each cell in the matrix such that only strong correlations between signals are further analyzed. While the threshold is chosen arbitrarily and may lead to the rejection of important correlations or the acceptance of weak ones, it is important to notice that correlations in electrical activity are ubiquitous in the brain and some filtering is needed for the data to be meaningful. Each correlation that passes the threshold is further treated as a functional link between their respective locations that become nodes (vertices), while all correlations that fail to meet the threshold are set to zero. In effect, a set of connections (edges) is obtained that can be portrayed as a network of functional connectivity.

Functional connectivity networks have been subjected to analysis using graph theoretical measures. Most of the research in this area focuses on determining stable properties of global functional networks of the brain in a resting state (that is, without engaging subjects in cognitive tasks). Just as has been found for anatomical connections (He et al. 2007; Sporns and Zwi 2004; Watts and Strogatz 1998), functional networks display small world properties but usually no scale-free nature (Salvador et al. 2005a; Salvador et al. 2005b; Yu et al. 2008). Interestingly, functional connectivity networks also display hierarchical modular organization (modules within modules), fairly similar across subjects (Meunier et al. 2009b), that is believed to be typical for complex systems and that enables them to be adaptive (Simon 1962). Such organization reflects the notion of functional units forming at different levels of description that enable smooth cognitive function.

A promising line of research has been devoted to differentiating the properties of functional networks in normal and pathological function of the brain – epilepsy (Ponten et al. 2007), Alzheimer's disease (Stam et al. 2007) and brain tumors (Bartolomei et al. 2006). In all these cases the pathology is related to a network structure closer to a random configuration of connection, while healthy subjects display small world properties in their functional connectivity networks. In contrast, healthy aging is associated with a functional network structure deviating from random connection configuration, preserving small world properties (Meunier et al. 2009a). However, older subjects have functional networks with fewer long-range, inter-modular connections and more short-range, intra-modular connections than younger subjects.

A few studies have investigated the relation of functional network measures and motor and cognitive tasks. Micheloyannis et al. (2006) compared structural properties of functional networks of groups of subjects differing in the number of years of formal education. The group that had only a few years of education had a more small-world like structure of their functional connectivity during a working memory task than the subjects with university degrees. The authors conclude that the more small-world network structure, being more functional, serves to increase the efficiency of processing that is specifically crucial for subjects with fewer years of education.

Bassett, Meyer-Lindenberg et al. (2006) have shown that whole brain functional connectivity networks in both resting state and motor task demonstrate fractal small-world properties, i.e. the small world structure is conserved across many frequency and topological scales. Moreover, the motor task increases the number of long-range connections between frontal and parietal cortices while still preserving the small world properties.

Finally, in a study of MEG derived functional networks, Valencia et al. (2008) have shown that a variety of network measures dynamically changes in a visual stimulus paradigm, evolving to meet the demands of the task. Namely, for the

15–25 Hz functional connectivity networks there is a peak in the clustering index, mean degree and efficiency at around 250 ms after the presentation of the stimulus. While there are changes in the network structure, the global small world properties are preserved. The authors conclude that such dynamic reconfiguration of the functional network serves efficient local and global functional integration.

Summary and Discussion

Cognitive function requires both segregation and integration of information, which imposes very demanding constraints on the system that produces it – the brain. Moreover, this segregation and integration of information needs to be structured in a hierarchical manner, with parts at one level being integrated into wholes that form units at the next level, while preserving their distinct, segregated features. Such hierarchical modular systems are believed to display complex properties and enable adaptation to a changing environment.

While the subjective experience of cognition is characterized by fluency and ease, it has so far proven very difficult to explain its neural basis. Our knowledge of the way information is segregated in the brain is by now sufficient – especially at the early stages of processing; however, we still lack definite understanding of how it is integrated into small and bigger functional units. The functional connectivity approach is a promising line of research that has the potential to explain how information may be locally and globally integrated while remaining segregated throughout the brain. By deriving its assumptions from the neural synchrony theories, it offers a comprehensive description of information processing during cognitive function.

First, it proposes a molecular mechanism for micro-integration – intermittent changes of synaptic strengths enable easy formation of synchronized cell ensembles. These produce a salient signal that can be distinguished – again, thanks to molecular mechanisms – from unsynchronized activity. Further, empirical data on synchronous oscillations recorded from larger regions points to their vital role in a variety of motor and cognitive functions, specifically in conscious processing. Finally, functional connectivity networks extracted from correlated signals permit a formal analysis of functionally integrated units to be made at various levels of description. The empirical results of such studies show how the global network properties relate to cognitive function as well as being able to distinguish between normal and pathological brain processing.

In summary, the concept of functional connectivity fulfills the requirements for explaining the neural correlates of cognition. While it is still a novel approach, it has a potential to finally provide us with a breakthrough in our understanding of the biological basis of the mind.

References

Aertsen, A.M., Gerstein, G.L., Habib, M.K., Palm, G.: Dynamics of neuronal firing correlation: modulation of "effective connectivity". J. Neurophysiol. **61**(5), 900–917 (1989)

Bartolomei, F., Bosma, I., Klein, M., Baayen, J.C., Reijneveld, J.C., Postma, T.J., Heimans, J.J., et al.: Disturbed functional connectivity in brain tumour patients: evaluation by graph analysis of synchronization matrices. Clin. Neurophysiol. **117**(9), 2039–2049 (2006). doi:10.1016/j. clinph.2006.05.018

Bassett, D.S., Meyer-Lindenberg, A., Achard, S., Duke, T., Bullmore, E.: Adaptive reconfiguration of fractal small-world human brain functional networks. Proc. Natl. Acad. Sci. **103**(51), 19518–19523 (2006). doi:10.1073/pnas.0606005103

Cajal, S.R.Y., Azoulay, L.: Histologie du système nerveux de l'homme & des vertébrés: avec fig. Maloine (1911)

Csépe, V., Juckel, G., Molnár, M., Karmos, G.: Stimulus-related oscillatory responses in the auditory cortex of cats. In: Pantev, W.C., Elbert, T., Lütkenhöner, B. (eds.) Oscillatory Event Related Brain Dynamics. Plenum Press, New York (1994)

Desmedt, J.E., Tomberg, C.: Transient phase-locking of 40 Hz electrical oscillations in prefrontal and parietal human cortex reflects the process of conscious somatic perception. Neurosci. Lett. **168**(1–2), 126–129 (1994). doi:10.1016/0304-3940(94)90432-4

Eckhorn, R.: Oscillatory and non-oscillatory synchronizations in the visual cortex and their possible roles in associations of visual features. Prog. Brain Res. **102**, 405–426 (1994)

Engel, A.K., König, P., Kreiter, A.K., Singer, W.: Interhemispheric synchronization of oscillatory neuronal responses in cat visual cortex. Science **252**(5010), 1177–1179 (1991)

Fell, J., Klaver, P., Lehnertz, K., Grunwald, T., Schaller, C., Elger, C.E., Fernandez, G.: Human memory formation is accompanied by rhinal-hippocampal coupling and decoupling. Nat. Neurosci. **4**(12), 1259–1264 (2001). doi:10.1038/nn759

Frien, A., Eckhorn, R., Bauer, R., Woelbern, T., Kehr, H.: Stimulus-specific fast oscillations at zero phase between visual areas V1 and V2 of awake monkey. Neuroreport **5**(17), 2273–2277 (1994)

Fries, P., Schroder, J.-H., Roelfsema, P.R., Singer, W., Engel, A.K.: Oscillatory neuronal synchronization in primary visual cortex as a correlate of stimulus selection. J. Neurosci. **22**(9), 3739–3754 (2002). doi:20026318

Gray, C.M., König, P., Engel, A.K., Singer, W.: Oscillatory responses in cat visual cortex exhibit inter-columnar synchronization which reflects global stimulus properties. Nature **338**(6213), 334–337 (1989). doi:10.1038/338334a0

He, Y., Chen, Z.J., Evans, A.C.: Small-world anatomical networks in the human brain revealed by cortical thickness from MRI. Cereb. Cortex **17**(10), 2407–2419 (2007). doi:10.1093/cercor/ bhl149

Hebb, D.O.: The Organization of Behavior. Wiley, New York/London (1949)

Hubel, D.H., Wiesel, T.N.: Receptive fields and functional architecture of monkey striate cortex. J. Physiol. **195**(1), 215–243 (1968)

Kohler, W.: Gestalt Psychology. Liveright, New York (1929)

Meunier, D., Achard, S., Morcom, A., Bullmore, E. (eds.): Age-related changes in modular organization of human brain functional networks. NeuroImage, **44**(3), 715–723 (2009). doi:10.1016/j.neuroimage.2008.09.062

Meunier, D., Lambiotte, R., Fornito, A., Ersche, K.D., Bullmore, E.T.: Hierarchical modularity in human brain functional networks. Front. Neuroinf. **3**(37) (2009b). doi:10.3389/ neuro.11.037.2009

Micheloyannis, S., Pachou, E., Stam, C.J., Vourkas, M., Erimaki, S., Tsirka, V.: Using graph theoretical analysis of multi channel EEG to evaluate the neural efficiency hypothesis. Neurosci. Lett. **402**(3), 273–277 (2006). doi:10.1016/j.neulet.2006.04.006

Miltner, W.H., Braun, C., Arnold, M., Witte, H., Taub, E.: Coherence of gamma-band EEG activity as a basis for associative learning. Nature **397**(6718), 434–436 (1999). doi:10.1038/17126

Murthy, V.N., Fetz, E.E.: Coherent 25–35-Hz oscillations in the sensorimotor cortex of awake behaving monkeys. Proc. Natl. Acad. Sci. U.S.A. **89**(12), 5670–5674 (1992)

Murthy, V.N., Aoki, F., Fetz, E.E.: Synchronous oscillations in sensorimotor cortex of awake monkeys and humans. In: Pantev, W.C., Elbert, T., Lütkenhöner, B. (eds.) Oscillatory Event Related Brain Dynamics, pp. 343–356. Plenum Press, New York (1994)

Pantev, C., Elbert, T.: The transient auditory evoked gamma-band field. In: Pantev, W.C., Elbert, T., Lütkenhöner, B. (eds.) Oscillatory Event-Related Brain Dynamics, pp. 219–230. Plenum Press, New York (1994)

Ponten, S.C., Bartolomei, F., Stam, C.J.: Small-world networks and epilepsy: graph theoretical analysis of intracerebrally recorded mesial temporal lobe seizures. Clin. Neurophysiol. **118**(4), 918–927 (2007). doi:10.1016/j.clinph.2006.12.002

Pulvermüller, F., Preißl, H., Eulitz, C., Pantev, C., Lutzenberger, W., Feige, B., Elbert, T., et al.: Gamma-band responses reflect word/pseudoword processing. In: Pantev, W.C., Elbert, T., Lütkenhöner, B. (eds.) Oscillatory Event Related Brain Dynamics, pp. 243–258. Plenum Press, New York (1994)

Rodriguez, E., George, N., Lachaux, J.P., Martinerie, J., Renault, B., Varela, F.J.: Perception's shadow: long-distance synchronization of human brain activity. Nature **397**(6718), 430–433 (1999). doi:10.1038/17120

Salvador, R., Suckling, J., Coleman, M.R., Pickard, J.D., Menon, D., Bullmore, E.: Neurophysiological architecture of functional magnetic resonance images of human brain. Cereb. Cortex **15**(9), 1332–1342 (2005a). doi:10.1093/cercor/bhi016

Salvador, R., Suckling, J., Schwarzbauer, C., Bullmore, E. (eds.): Undirected graphs of frequency-dependent functional connectivity in whole brain networks. Philos. Trans. Roy. Soc. B: Biol. Sci. **360**(1457), 937–946 (2005). doi:10.1098/rstb.2005.1645

Simon, H.A.: The architecture of complexity. Proc. Am. Philos. Soc. **106**(6), 467–482 (1962)

Singer, W.: Neuronal synchrony: a versatile code for the definition of relations? Neuron 24(1), 49–65, 111–125 (1999)

Singer, W., Gray, C.M.: Visual feature integration and the temporal correlation hypothesis. Annu. Rev. Neurosci. **18**(1), 555–586 (1995). doi:10.1146/annurev.ne.18.030195.003011

Sporns, O., Zwi, J.: The small world of the cerebral cortex. Neuroinformatics **2**(2), 145–162 (2004). doi:10.1385/NI:2:2:145

Stam, C.J., Reijneveld, J.: Graph theoretical analysis of complex networks in the brain. Nonlinear Biomed. Phys. **1**(1), 3 (2007). doi:10.1186/1753-4631-1-3

Stam, C.J., Jones, B., Nolte, G., Breakspear, M., Scheltens, P.: Small-world networks and functional connectivity in Alzheimer's disease. Cereb. Cortex **17**(1), 92–99 (2007). doi:10.1093/cercor/bhj127

Stopfer, M., Bhagavan, S., Smith, B.H., Laurent, G.: Impaired odour discrimination on desynchronization of odour-encoding neural assemblies. Nature **390**(6655), 70–74 (1997). doi:10.1038/36335

Tallon-Baudry, C., Bertrand, O., Delpuech, C., Pernier, J.: Oscillatory gamma-band (30–70 Hz) activity induced by a visual search task in humans. J. Neurosci. **17**(2), 722–734 (1997)

Tononi, G., Srinivasan, R., Russell, D.P., Edelman, G.M.: Investigating neural correlates of conscious perception by frequency-tagged neuromagnetic responses. Proc. Natl. Acad. Sci. U.S.A. **95**(6), 3198–3203 (1998)

Valencia, M., Martinerie, J., Dupont, S., Chavez, M.: Dynamic small-world behavior in functional brain networks unveiled by an event-related networks approach. Phys. Rev. E **77**(5), 050905 (2008). doi:10.1103/PhysRevE.77.050905

von der Malsburg, C.: The correlation theory of brain function. In: Domany, E., Hemmen, J.L. (eds.) Models of Neural Networks II: Temporal Aspects of Coding and Information Processing in Biological Systems, Chapter 2, pp. 95–119. Springer, New York (1994) von Stein, A., Sarnthein, J.: Different frequencies for different scales of cortical integration: from local gamma to long range alpha/theta synchronization. Int. J. Psychophysiol. **38**(3), 301–313 (2000). doi:10.1016/S0167-8760(00)00172-0

von Stein, A., Rappelsberger, P., Sarnthein, J., Petsche, H.: Synchronization between temporal and parietal cortex during multimodal object processing in man. Cereb. Cortex **9**(2), 137–150 (1999). doi:10.1093/cercor/9.2.137

Watts, D.J., Strogatz, S.H.: Collective dynamics of 'small-world' networks. Nature **393**(6684), 440–442 (1998). doi:10.1038/30918

Wrobel, A., Ghazaryan, A., Bekisz, M., Bogdan, W., Kaminski, J.: Two streams of attention-dependent beta activity in the striate recipient zone of cat's lateral posterior-pulvinar complex. J. Neurosci. **27**(9), 2230–2240 (2007). doi:10.1523/JNEUROSCI.4004-06.2007

Yu, S., Huang, D., Singer, W., Nikolić, D.: A small world of neuronal synchrony. Cereb. Cortex **18**(12), 2891–2901 (2008). doi:10.1093/cercor/bhn047

Chapter 3
A Dynamical Systems Approach to Conceptualizing and Investigating the Self

Urszula Strawińska

Abstract This chapter presents the Self from the perspective of dynamical systems. Change and dynamics are integral aspects of the human experience; the dynamical systems perspective captures this essence of the Self. It has advanced our understanding of many self-related concepts that previously were difficult to address accurately within a non-dynamic framework. The focal point of this chapter is the Society of Self model that was inspired by the advances in nonlinear systems theory and uses such concepts as self-organization, emergence, and the notion of attractors in order to conceptualize how throughout their lives people build, develop, and use a well-integrated sense of who they are. The usefulness and unique contribution of the dynamical systems approach in psychology are illustrated with simulations and new empirical results from a variety of lines of research exploring the origin, functioning, and implications of self-related concepts and processes, defined in dynamical terms.

Introduction

The power of dynamical systems applications in psychology lies in their ability to provide meaningful conceptualization of important aspects of many psychological phenomena or specific aspects of those phenomena that previously were difficult to investigate and understand. As social scientists with the tools and methods rooted in dynamical systems theory we are better equipped for the fascinating journey into the world of group interactions, interpersonal love and hate relationships, and finally into the mind of an individual, where it all takes place.

One of the very interesting areas of scientific investigation in psychology, to which dynamical systems approach was applied, is the psychology of Self. The Self

U. Strawińska
Department of Psychology, Warsaw School of Social Sciences and Humanities,
Chodakowska 19/31, 03815 Warsaw, Poland
e-mail: ula.strawinska@gmail.com

A. Nowak et al. (eds.), *Complex Human Dynamics*, Understanding Complex Systems,
DOI 10.1007/978-3-642-31436-0_3, © Springer-Verlag Berlin Heidelberg 2013

is a structure that has a very particular status in the cognitive system of every one of us. It is larger and more complex than any other structure (e.g. Klein and Loftus 1988; Markus 1983; Rogers et al. 1977) because it holds enormous amount of information about ourselves and about all that has happened in our lives and that we consider in any way, even ever so slightly, to be personally pertinent. Our memories, thoughts and feelings, values, goals and fears, no matter whether real or imagined, desired or avoided, as well as all kinds of self-relevant information appear in the stream of consciousness (James 1890) when we interact with the outside world and when we self-reflect. Moreover, "thoughts about oneself are arguably the most recurrent and salient elements in the stream of consciousness" (cf. Vallacher and Nowak 2000, p. 35). It is truly remarkable that, despite the vast amount of knowledge already internalized and a continuous exposure to information relevant to self-understanding present in the outside world, people are able to build and maintain a coherent picture of themselves. How does the self-concept emerge from our life long experience? How do we decide which features define the real 'me', describing who we really are? How do we manage to filter incoming information so that our sense of who we are is not disrupted by every-day ups and downs? Why are some of us more susceptible to what other people think of us and allow criticism to easily undermine our feeling of self-worth? How do people manage to find a balance between knowing who they are and staying open to potentially adaptive changes in what they think about themselves?

Approaching these fundamental and long-standing questions in self-theory from the perspective of dynamical systems laid the grounds for dynamical theories of the Self and shed a new light on a variety of crucial self-related psychological constructs (e.g. self-esteem) and processes (e.g. self-esteem maintenance). The purpose of this chapter is to familiarize the reader with applications of concepts and tools stemming from dynamical systems theory in the domain of Self psychology. The focal point of the first part of the chapter is the dynamical model called Society of Self (Nowak et al. 2000). This model describes the emergence and maintenance of evaluative differentiation of the self-concept as a result of self-organizing processes among individual pieces of self-relevant information, while the notion of attractors, discussed in the second part of the chapter, provides the mechanisms that link organization of the self-concept to the origin of subjective experience of stability and continuity in self-understanding. Together these two applications of dynamical systems to the theory and research on Self present a coherent and convincing example that dynamical conceptualization of self-related phenomena does indeed capture the essence of the Self.

Society of Self Model: Self as a Complex Dynamic System

In 2000 Nowak and coworkers (Nowak et al. 2000) presented a dynamic model of Self, called Society of Self, that was inspired by a novel cross-disciplinary approach. Their model is based on the idea that systems at different levels of analysis that come from domains as disparate as physics, economics, or the social world, display formal similarities that make it possible to analyze their structure and dynamics with the use of tools and methods drawing on advances in research on nonlinear dynamical systems.

The Society of Self model conceptualizes Self as a nonlinear dynamical system that, by definition, is a set of interconnected elements that evolve over time (Vallacher and Nowak 2000, p. 36). The elementary constituents of the self-system are the cognitive representations of specific pieces of self-relevant information. These elements belong to a variety of cognitive domains (e.g. self-perceived traits, physical features, memories) and reflect the richness of individual experiences. The variety in content that they present is potentially unlimited. They all have, however, a common 'denominator': they all carry an evaluative signature. In other words, every piece of self-relevant information can be assigned valance on a continuum ranging from negative to positive. For instance, the awareness of one's Achilles heel is most likely evaluated negatively, while pride in one's accomplishment or talents carries positive valance.

Thanks to the existence of a common dimension of evaluation, individual pieces of information can be scaled, compared, weighted and, as a result, integrated into more abstract cognitive structures. For instance, positive evaluation of one's cooking skills, punctuality, and sense of humor, allows for these elements to become a foundation of an evaluatively coherent construct on a higher level, which might take on a form of identifying oneself as a good parent.

Importantly, the Society of Self model assumes that the valance of individual elements residing in the self-structure is not fixed but to a large extent depends on the context. Certain pieces of information might seem positive or negative, depending on the bigger picture that emerges. Consider this example. How would you feel about a driver whom you watched speeding, going through red lights, changing lanes rapidly? Reckless? Careless? Adrenaline seeker? Bad driver? What if you found out that he just received a call from his parent who needed urgent hospitalization and thus was rushing home to the rescue? Would that change your evaluation? In a similar way the evaluation of self-relevant information is subject to change as the frame of reference provided by thematically relevant elements changes. To the extent that it is feasible, individual pieces of self-relevant information influence each other. In particular, they change their own valance or change the valance of the related elements to form evaluatively coherent units on a higher level. The processes that lead to changes in evaluation of individual elements can take on a form of one of the self-defense mechanisms well-established in the literature (e.g. Tesser 2001; Tesser et al. 2000). Such mechanisms as, for example, denial, selective recall, or dissonance reduction are all aimed at addressing a detected incongruity in processed information that can potentially threaten the integrity of the existing self-concept and for that reason it triggers some cognitive operations that will protect their evaluative coherence.

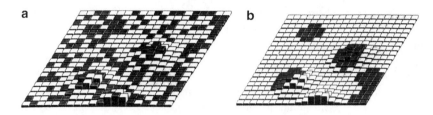

Fig. 3.1 Computer simulation of self-organization process. Each tile represents individual pieces of self-knowledge. *Light colored* tiles have positive evaluation, *dark colored* tiles have negative evaluation. The height of each tile is proportional to the relative importance of each element. The distance between the elements shows how individual elements are related. (**a**) Shows an undifferentiated system. (**b**) Shows a differentiated system (Nowak et al. 2002)

Evaluative Differentiation in Self-Concept Through Self-Organization

As a result of mutual influences between individual elements, information in the self-system becomes differentiated into evaluatively coherent subareas (Showers 1992). To illustrate this process of evaluative differentiation of a self-system with properties defined by the Society of Self model, Nowak et al. (2000) implemented computer simulations. They used a cellular automata model, depicted on Fig. 3.1a, b, in which each tile on a two-dimensional grid represents an individual element of self-knowledge (e.g. self-pertinent memory). In addition to carrying evaluative signature (positive evaluation denoted by light colored tiles on the grid, or negative evaluation denoted by dark color) each element is also characterized by its relative importance (also conceptualized as centrality) for self-understanding, which in the model is denoted by the height of the tiles (importance and height are positively correlated). More central aspects of the Self have a greater chance of influencing the evaluation of neighboring elements, and also they are more resistant to changing their own evaluative state under the influence of their neighbors.

Simulation results showed that when individual elements of a system adjust their binary evaluative state to the valance that prevails among the surrounding elements at a given moment, an entire system with initially random configuration of positive and negative components (see Fig. 3.1a) is transformed into a system composed of a number of larger areas with shared evaluation (see Fig. 3.1b).

The temporal evolution of this model demonstrates the creation of evaluatively coherent substructures resulting in evaluative differentiation of the self-concept. The evaluative differentiation visible on the level of the system is not related to the operation of any higher order supervisory mechanism but results from rules that dictate behavior of individual elements (Vallacher and Nowak 2000). What these local interactions will eventually produce depends on the existing structure of the system (how the constituent parts are organized) and on the nature of influences between the elements (due to the press for integration individual elements try to achieve evaluative congruence by changing their evaluative state). Therefore, the resulting evaluative landscape (see Fig. 3.1b) is said to emerge from the self-organizing processes taking place on a low level in the entire system. Evaluative differentiation of the self-concept emerges through self-organization of pieces of self-knowledge, triggered by the press for evaluative integration. (For a more detailed description of the cellular automata model the reader is encouraged to refer to the original article by Nowak et al. (2000) and to the cellular automata model of social influence (Nowak et al. 1990).)

The Society of Self model proposes that self-organization is the mechanism underlying the organization of self-relevant information in the self-concept. The evaluatively congruent higher-order substructures, that are created through the process of evaluative differentiation, correspond to social roles, areas of competence, or, for example, behavioral standards. From isolated instances of their experience a person may, for example, create a positive view of themselves as a student, a moderately favorable self-view as a musician, and a self-perception of themselves as a poor public speaker. To the extent that incongruent elements allow for reinterpretation, reevaluation, or discounting during the formative stage of a self-perception, a self-perception in question will take on an unequivocal evaluation. By the same token, an evaluatively incoherent self-view results from the presence of a portion of contradictory information that is resistant to change due to its importance, salience, or indisputability. The level of agreement between evaluations of elements within a given self-aspect is captured by a construct of evaluative coherence that is manifested on a phenomenological level as certainty. In an empirical study McMillan (2003, 2005) showed that evaluative coherence and subjective experience of certainty are correlated. In this study participants filled out questionnaires that measured their evaluation with respect to a set of roles (e.g. student) and traits (e.g. punctual), as well as certainty (e.g. How certain are you about your self-evaluation as a student?). Evaluative coherence of the self-evaluation was also assessed with the use of a questionnaire where specific self-aspects were presented as a continuum ranging from negative to positive evaluation (e.g. lazy to hardworking). Based on participant's responses the degree of evaluative coherence was estimated. Results revealed that when participants described themselves with respect to self-aspects that were evaluatively coherent they also reported more certainty concerning their standing on those particular dimensions. When they characterized themselves with respect to less evaluatively coherent self-aspects they lacked certainty in self-description. A lack of evaluative congruence between pieces of information causes an individual, who is trying to use this information to arrive at conclusions about that aspect of self-perception, to undermine their confidence or, in the case of an evident and large contradiction, makes any generalization impossible to form. For example, both spectacular successes and shameful failures at performing a certain task, let us think about cooking, make it difficult to form an answer to whether we are good at cooking or not. In the face of contradictory information our self-view as a cook would lack definite evaluation and, as a result, lack certainty. To arrive at a definite opinion we would seek more evidence, e.g. more feedback from our guests, or we could try to change the level of identification (Vallacher and Wegner 1987), where, instead of struggling to form a global opinion about cooking skills, we try to look into finer distinctions between different types of dishes or regions of cuisine. Instead of a global view as a cook, more fine-grained pride of one's baking skills held with confidence could emerge.

The concept of evaluative coherence and a general idea of evaluative differentiation was investigated in a series of simulations and empirical studies conducted by Nowak and coworkers. One line of investigation looked into implications of evaluative organization of the self-concept for the effectiveness of regulation processes in which the Self is involved. An evaluatively differentiated self-concept, as opposed to a self-concept with mixed organization of positive and negative elements, is more adaptive (e.g. Linville 1985, 1987; Nowak et al. 2000; Showers and Zeigler-Hill 2007; Showers et al. 2006). To the extent that people are able to build coherently evaluated areas that define their strengths and weaknesses, which is a basis for better and more stable self-understanding and gives a very much desired sense of personal integration and self-certainty, they can then use those well-defined self-aspects as guidance for the regulation of emotions, thought processes, and actions. This regulating function of the self-concept can be illustrated through results coming from a line of simulation studies that explored the behavior of the self-system in response to external influences. The overall picture that emerged, corroborates the hypothesis that evaluative differentiation in the self-system plays an important role in maintaining its relative stability in the face of incoming information. Because of the organization of the elements into coherent substructures, individual pieces of self-knowledge provide mutual support for the evaluation they share with their neighbors. As a result, incoming information with a conflicting evaluation has a lesser potential to exert an influence on the existing self-view. When, for example, social feedback brings criticism with respect to a self-aspect that is unequivocally defined as positive, this self-perception is not very likely to change in response to new information, unless the salience of the negative remark is truly overwhelming. By the same token, due to the lack of mutual support coming from individual elements, an incoherent self-aspect is open to redefinition and change induced by new contradictory information pertinent to the domain in question. To provide empirical verification of the simulation results, Strawińska (2006) conducted a study that supported the finding that reaction to incoming self-relevant information depends on the level of evaluative coherence of a given self-aspect. In the study participants' self-perceptions were investigated and evaluative coherence of the dimension of consciousness was captured in a self-report questionnaire. Participants were subsequently confronted with a bogus feedback, allegedly created by the partner of interaction. In reality the feedback was based on each participants' self-perceptions and either reflected them or contradicted them. For example, in the confirmatory condition a conscientious person found out that they were perceived as conscientious, while in the disconfirmatory condition a conscientious person would be described as unconscientious. The analysis focused on how participants perceived the feedback presented to them. Specifically, they were asked to assess how accurate this information was. In support of the study's hypothesis, the assessment of feedback was contingent upon the level of evaluative coherence of participants' self-view.

Participants with a coherent self-view rated confirmatory self-relevant information as accurate and rejected disconfirmatory information as inaccurate. In contrast, individuals with incoherent self-views did not differentiate between confirmatory and disconfirmatory feedback in their accuracy ratings. The hedonistic tone of the feedback did not matter as much as the evaluative coherence of the self-view. For example, negative feedback was perceived as accurate by individuals who had a well defined negatively valenced self-perception in the domain of conscientiousness. These results suggest that the evaluative coherence of a self-view is advantageous for the effective operation of processes responsible for selective acceptance and rejection of self-relevant information. Response to social feedback is one important aspect of regulatory functions that the Self performs. The ability to refer to a firm set of knowledge about mastered skills, self-perceived traits, important personal values, and preferences gives people a stable frame of reference necessary to make favorable decisions about in what to invest the resources, with whom to make friendships, and about a whole range of every day more or less significant choices. Research results suggest that the same processes that produce an evaluatively differentiated self-concept (i.e. press for evaluative integration among elements residing in the self-concept operating through self-organization) allow the Self to protect its integrity and respond adaptively to incoming information.

Attractors as a Source of Stability in Self-Understanding

To fully appreciate the insights provided by the dynamical approach, we take a closer look at the concept of attractors which explains how stability emerges from interactions of structure and dynamism in the Self system. Throughout their lives people strive to maintain integration in how they view themselves. Coherence and continuity in self-understanding allow them to make long-term plans, arrive at optimal decisions, and perform complex individual and collective actions. Despite the myriad of factors (e.g. compliments, letdowns) that could potentially change their opinion about themselves from moment to moment, what they think, and more importantly, how they feel about themselves, achieves sufficient constancy for such global variables as self-esteem, and, more recently, self-esteem stability to become the subject of scientific investigation in psychology theory and research, as well as the subject of a wide public interest (e.g. Rosenberg 1965; Showers et al. 2006). The concept of self-esteem captures a generalized subjective feeling of self-worth. The implications of high and low self-esteem are remarkable. High self-esteem is linked to benefits that fall into two broader categories of enhanced initiative (e.g. willingness to speak up in groups) and pleasant feelings (e.g. happiness). Low-self esteem, under some circumstances, may lead to depression (for a critical review of benefits and downsides of high and low self-esteem see e.g. Baumeister et al. 2003). Despite far-reaching implications, the nature of the construct itself is still a subject of an intense debate.

From the standpoint of dynamical systems, self-esteem and stability of self-esteem are considered emergent properties of the self-system. It is not the sheer amount of positive or negative content of the self-concept that determines either of them. In fact, people with high global self-esteem tend to have a more evaluatively differentiated self-concept, whereas those that score lower on self-esteem have a more evaluatively mixed self-concept, which implies that low self-esteem does not derive from well articulated negative convictions about the Self but rather results from the presence of both negative and positive, yet evaluatively disintegrated, self-views. The processes underlying the emergence of self-esteem and self-esteem stability do involve the content, but also the structure, the dynamism, and interactions of those factors. The structure of the self-concept, that is how the self-knowledge components are organized in the self-concept, guides self-reflection. "In an unperturbed system, the flow of self-reflective thought tends to move from regions with low coherence values toward those with higher values" (Vallacher et al. 2002). The trajectory of self-reflection in turn underlies the stability of self-evaluation over time. The attractor dynamics accurately captures how self-reflection operates on an evaluatively differentiated self-concept.

The Concept of Attractor

An attractor is a state toward which a system evolves over time. In other words, it is a state preferred by the system. When the system is at its attractor state, this state is fairly stable; it does not change, unless external influences perturb the system. In case of sufficiently strong perturbations, the system will move away from the attractor state toward a state induced by the nature of the external influence, but it will have a tendency to return to its attractor when external forces abate. There can be more than one attractor in a system. Attractors can be characterized with respect to two basic features, their strength reflected in the basin of attraction and in their resistance to external influences. The width of the basin of attraction is related to the number and variety of states that lead toward this attractor. The resistance to external influences corresponds to how easy it is to perturb system so that it moves away from the attractor state. There are three major types of attractors that are related to three distinct patterns underlying the temporal evolution of a system. When a system evolves toward a stable equilibrium, this denotes that a fixed-point attractor is present. When states of a system alternate between different states, a periodic attractor might be present. An attractor in the form of a chaotic attractor underlies a seemingly random behavior of the system, which however is deterministic, and therefore allows for prediction. For the conceptualization of the emergence of global properties of the Self the concept of fixed point attractors is most relevant.

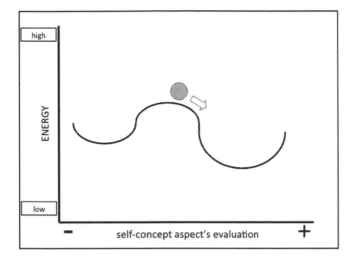

Fig. 3.2 Energy landscape depicting self-concept with two attractors and positive inclination

Attractors in the Self-System

Evaluatively coherent areas of the Self function as attractors for self-reflection. Thoughts that appear during an uninterrupted self-reflection belong to a self-area characterized by evaluative congruence. When the self-reflection activates positive thoughts, other positive self-aspects are more likely to appear in the stream of consciousness than negative thoughts, and vice versa. The influence of external factors is constrained by the presence of an attractor and is determined by the combined effect of internal characteristics of the system (i.e. number, strength, and resistance of attractors) and the nature of that influence (i.e. contradictory, neutral, confirmatory information) as well as its strength (e.g. influence that comes from an important source, i.e. a significant other, has a greater potential to induce a change than influence coming from someone who is not important to the actor).

Figure 3.2 shows a differentiated self-concept as denoted by the presence of two regions with attracting tendencies, in the left (negatively valenced) and right (positively valenced) part of the picture. This is a self-concept with a positive bias suggested by the relatively deeper attractor in the positive region. Due to the working of the attractors in this system, this system tends to stabilize in the vicinity of the positive attractor. When perturbed by external forces, it may switch to the negative attractor; however because this attractor is more shallow than the positive attractor the system is likely to return to the positive state over time. To bring this reasoning on the level of self-evaluation, a person whose self-concept could be illustrated with this energy landscape would most of the time have positive self-evaluation. In the face of failures, critical remarks or any other event that has negative implications for their self-view, the self-evaluation might become less positive, but it will return to its

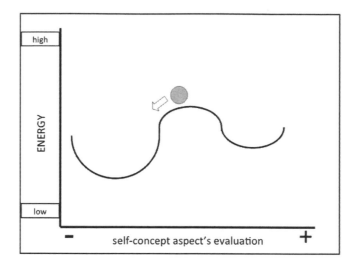

Fig. 3.3 Energy landscape depicting self-concept with two attractors and negative inclination

dominant attractor state with time. It is easy at this point to think of an opposite scenario for someone whose dominating (or the only) attractor resides in the range of negative evaluation (see Fig. 3.3). However, most healthy individuals have a positively biased self-concept (Taylor and Brown 1988). Negative self-evaluation linked to depressive states is related, not so much to the existence of a strong attractor in negative areas of self-evaluation, but to a lack of any attractors that would stabilize self-evaluation (Johnson and Nowak 2002).

A different scenario for the trajectory of self-evaluation captures the functioning of a self-system with two strong attractors residing in areas of negative and positive evaluation, shown in Fig. 3.4. Such a system displays bi-stable behavior. Once in the vicinity of the positive attractor, self-evaluation stabilizes in positive regions. However, when strong enough influences push the system out of the basin of positive attraction, it will fall into a negative state and remain there until something similar with respect to the strength but opposite in the implications for self-evaluation happens.

As an illustration of the interplay of structure and dynamism, the metaphor of landscapes of self-reflection (cf. Vallacher and Nowak 2000) comes to mind. This metaphor is illustrated on Figs. 3.2, 3.3 and 3.4. An energy landscape is created to represent the organization of the self-concept. In this landscape hills correspond to high energy, valleys correspond to low energy. The stream of self-reflection has the tendency to go to low energy areas, i.e. valleys, and to escape from high energy areas, i.e. hills. This property is illustrated by the ball that represents a current energy state and tends to roll down the hills and rest in valleys in search of a local energy minimum. The energy level of a given area denotes the degree of evaluative consistency (the greater the evaluative coherence, the lower the level of energy).

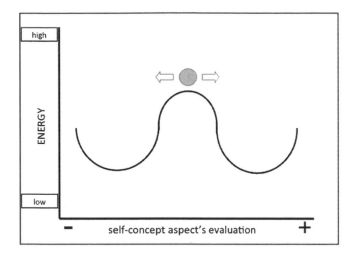

Fig. 3.4 Energy landscape depicting self-concept with two strong attractors in the negative and positive region

In other words, areas of low energy serve as attractors for the stream of self-reflection. When the self-reflection activates areas with coherent evaluation, the self-evaluation remains stable. If conscious attention is directed to incoherent areas, the self-evaluation fluctuates and reflects the variability in evaluation of individual elements in that area of the self-concept. "In a region of self-structure marked by a high degree of evaluative consistency, the energy value is close to 0, signifying that all the cognitive elements in this region share a common valence. (. . .) In a region marked by weaker evaluative consistency, the energy is correspondingly greater in magnitude, with self-reflection exhibiting some variability in self-evaluation over time" (cf. Vallacher and Nowak 2000, pp. 45–46).

Attractor dynamics is very useful in explaining the link between structure and dynamism that produces stability visible on the level of global parameters characterizing the self-system. The regularities visible in the stream of self-reflection follow certain temporal patterns, that can be predicted and investigated with more precision when coherent regions of the Self are conceptualized as attractors in the self-system.

Closing Remarks

The dynamical perspective resonates well with the currently widely shared view of the Self that defines it as a dynamic, multifaceted knowledge structure (e.g., Greenwald and Pratkanis 1984; Kihlstrom and Cantor 1984; Markus and Wurf

1987). It is not an individual thought or deed but a number of conglomerates of self-attributes immersed and anchored in the context of our idiosyncratic experience that define who we are in our own eyes. Organizing knowledge about ourselves with respect to how positively vs. negatively it reflects on us creates coherent self-views that become the building blocks of our identity and provide a frame of reference for our thoughts, feelings, and actions.

This dynamical model of Self provides an interesting framework within which the seemingly paradoxical phenomena of malleability and constancy of the Self no longer seem mutually exclusive, but are reconciled to gain more trustworthiness in expressing the essence of the Self and self-related processes. The Self is both static, and dynamic, stable, yet responsive to the environment, firm and adapting. Thanks to that this largest and most important cognitive structure is able to generalize the wealth of individual experience in a meaningful way and help individuals navigate in the complex social and physical environment.

The dynamical systems approach adheres to the principals of dynamical minimalism (Nowak 2004). It treats parsimony as the guidance in theory construction, but it retains depth and richness in understanding. It is through the use of such concepts as self-organization, emergence, or attractor dynamics that the dynamical conceptualization of psychological phenomena is successful at providing a fairly simple, yet valid explanation with due acknowledgment of their complexity. "In dynamical models, simple rules on the lower level can lead to the emergence of very complex structures and processes an [sic] the higher level. Thus, without forfeiting our depth of understanding regarding the phenomena, we can propose simple rules that will generate the phenomenon in its complexity" (cf. Nowak 2004, p. 190).

References

Baumeister, R.F., Campbell, J.D., Krueger, J.I., Vohs, K.E.: Does high self-esteem cause better performance, interpersonal success, happiness, or healthier lifestyles? Psychol. Sci. Publ. Inter. **4**, 1–44 (2003)

Greenwald, A.G., Pratkanis, A.R.: The self. In: Wyer, R.S., Srull, T.K. (eds.) Handbook of Social Cognition, pp. 129–178. Erlbaum, Hillsdale (1984)

James, W.: The Principles of Psychology. Holt, New York (1890)

Johnson, S.L., Nowak, A.: Dynamical patterns in bipolar depression. Pers. Soc. Psychol. Rev. **6**, 380–387 (2002)

Kihlstrom, J.E., Cantor, N.: Mental representations of the self. In: Berkowitz, L. (ed.) Advances in Experimental Social Psychology, vol. 15, pp. 1–47. Academic, New York (1984)

Klein, S.B., Loftus, J.: The nature of self-referent encoding: the contributions of elaborative and organizational processes. J. Pers. Soc. Psychol. **55**, 5–11 (1988)

Linville, P.W.: Self-complexity and affective extremity: don't put all of your eggs in one cognitive basket. Soc. Cogn. **3**, 94–120 (1985)

Linville, P.W.: Self-complexity as a cognitive buffer against stress-related illness and depression. J. Pers. Soc. Psychol. **52**, 663–676 (1987)

Markus, H.: Self-knowledge: an expanded view. J. Pers. **51**, 543–565 (1983)

Markus, H., Wurf, E.: The dynamic self-concept: a social psychological perspective. Annu. Rev. Psychol. **38**, 299–337 (1987)

McMillan, K.K.: Dynamics of self-system coherence: relations of entropy and global properties. Master's thesis, Florida Atlantic University, 2003. Masters Abstr. Int. **41**, 1529 (2003)

McMillan, K.K.: Coherence versus fragmentation in the self-system: implications for self-evaluation and social behavior. Doctoral dissertation, Florida Atlantic University, 2005. Dissertation Abst. Int. **66**, 1222 (2005)

Nowak, A. Dynamical minimalism: Why less is more in psychology. J. Pers. Soc. Psychol., 8,183–192 (2004)

Nowak, A., Szamrej, J., Latané, B.: From private attitude to public opinion: a dynamic theory of social impact. Psychol. Rev. **97**, 362–376 (1990)

Nowak, A., Vallacher, R.R., Tesser, A., Borkowski, W.: Society of self: the emergence of collective properties in self-structure. Psychol. Rev. **107**, 39–61 (2000)

Rogers, T.B., Kuiper, N.A., Kirker, W.S.: Self-reference and the encoding of personal information. J. Pers. Soc. Psychol. **35**, 677–688 (1977)

Rosenberg, M.: Society and the Adolescent Self-Image. Princeton University Press, Princeton (1965)

Showers, C.: Compartmentalization of positive and negative self-knowledge: keeping bad apples out of the bunch. J. Pers. Soc. Psychol. **62**, 1036–1049 (1992)

Showers, C.J., Zeigler-Hill, V.: Self-structure and self-esteem stability: the hidden vulnerability of compartmentalization. Pers. Soc. Psychol. Bull. **33**, 143–159 (2007)

Showers, C.J., Zeigler-Hill, V., Limke, A.: Self-structure and emotional maltreatment: successful compartmentalization and the struggle of integration. J. Soc. Clin. Psychol. **25**, 473–507 (2006)

Strawińska, U.: Evaluative coherence of the self-concept and cognitive and affective processing of self-relevant information. Master's thesis, Warsaw School of Social Sciences and Humanities, Warsaw, Poland (2006)

Taylor, S.E., Brown, J.D.: Illusion and well being: some social psychological contributions to a theory of mental health. Psychol. Bull. **103**, 193–210 (1988)

Tesser, A.: On the plasticity of self defence. Curr. Dir. Psychol. Sci. **10**, 66–69 (2001)

Tesser, A., Crepaz, N., Collins, J.C., Cornell, D., Beach, S.R.H.: Confluence of self defense mechanisms: on integrating the self zoo. Pers. Soc. Psychol. Bull. **26**, 1476–1489 (2000)

Vallacher, R.R., Nowak, A.: Landscapes of self-reflection: mapping the peaks and valleys of self-evaluation. In: Tesser, A., Felson, R., Suls, J. (eds.) Psychological Perspectives on Self and Identity, pp. 35–65. American Psychological Association, Washington, DC (2000)

Vallacher, R.R., Wegner, D.M.: What do people think they are doing? action identification and human behavior. Psychol. Rev. **94**, 3–15 (1987)

Vallacher, R.R., Nowak, A., Froelich, M., Rockloff, M.: The dynamics of self-evalutaion. Pers. Soc. Psychol. Rev. **6**, 370–379 (2002)

Chapter 4
Novelty Recognition Models

Michał Ziembowicz

Abstract In the following chapter a theoretical model of novelty recognition mechanism is described. In the first part the insights as well as shortcomings of classic approaches (e.g. mere exposure, feeling of knowing, tip of the tongue) are discussed. Those models are mostly descriptive and do not provide the explanatory mechanism for the observed effects. Such a mechanism is proposed in the second part of the chapter. First a metacognitive mechanism of fluency is described and identified as an empirical counterpart of a neural network measure of coherence. The application of a coherence/fluency-based feedback loop allows for building a precise and parsimonious model of novelty recognition process.

Introduction

Novelty recognition is one of the most basic cognitive functions. It is very fast and very imprecise, allowing for quick binary choices rather than complicated decision making. As such it may seem unimportant and obsolete – a remnant from early stages of evolution. Yet if we think about it for a while we will have to admit that the amount of fast zero–one choices made every day exceeds by far the number of complex decisions we have to undertake in the same period. Each time we look at somebody else's face we (or rather our brains) decide if this person is known to us; every time we return home we recognize the things we have left there when we were leaving and they feel familiar to us; each time we listen to a music tune we immediately know whether or not we knew it already and whether we liked it or detested it. More examples can easily be found. In fact almost everything we do or think of can be broken into series of simple binary divisions between known and unknown, liked and disliked etc.

M. Ziembowicz
Institute for Social Studies, University of Warsaw, Stawki 5/7, Warsaw 00183, Poland
e-mail: ziembowicz@gmail.com

A. Nowak et al. (eds.), *Complex Human Dynamics*, Understanding Complex Systems, 49
DOI 10.1007/978-3-642-31436-0_4, © Springer-Verlag Berlin Heidelberg 2013

One might ask why should it be so important to tell new from familiar objects (and it certainly is important since we possess specialized and automatic cognitive mechanism for it). The answer is quite simple: new or unexpected stimuli can be a sign of either danger or a sudden possibility of satisfying a need and therefore require a fast reaction. As Robert Zajonc (1984) said: things that are known must have been encountered before (possibly many times) and did not lead to any harm or loss of life, therefore can be considered safe; on the other hand novel objects are unpredictable and should be treated with caution, hence a person is alerted with negative affect. This is exactly why novel stimuli most strongly draw one's attention. This mechanism was described by Sokolov (1963) as the orienting reflex. When a novel stimulus appears it stops current activity and the cognitive resources are allocated to deal with it. One does not need to be explicitly aware of the fact that the perceived object is new. The cognitive system produces affective signals indicating the level of familiarity. In general familiarity is connected with positive reaction and novelty with negative.

The common sense explanation of the novelty recognition mechanism could be formulated as follows: first an object is perceived and then it is classified to one of the known categories. If the classification process does not succeed, it means that the stimulus is new. The novelty recognition is therefore the last in the sequence. The experimental observations, as well as everyday life experience, leads to a quite opposite conclusion. Not only are people able to tell new from known objects very fast, before they recognize them, but the whole process can happen even without awareness. This seemingly paradoxical situation has been shown many times with the subliminal mere exposure procedure, where people are presented with the visual stimuli exposed for a very short time (e.g. 2 ms) (see Bornstein 1989 for an extensive review). It is not possible for anybody to consciously perceive anything that lasts for this short a time. However the participants were later reacting differently to the stimuli that had been presented in such a way than to new stimuli that they had never seen before. The common sense explanation clearly cannot be applied here. Yet how is it possible to tell if something is 'known' before it was actually been recognized?

In this chapter various approaches to the novelty recognition will be presented, together with proposed explanations of the process. We will look at phenomena such as mere exposure and feeling of knowing where this mechanism plays a central role.

Mere Exposure

The mere exposure effect can be summarized in one sentence: repeated presentation of a stimulus, without any additional reinforcement, leads to a more positive reaction to this stimulus as compared to one not encountered before. It was described by Robert Zajonc in 1968, not for the first time though. Similar observations had been made many times before, from the first days of psychology.

Fechner (1876) and James (1890) wrote about the preference for the musical tunes that are heard often (it should be noted that the observation was made way before sound recording was possible). In social psychology it was known for a long time that we tend to like more those people we meet more often (Festinger 1950). The effect was also known for a very long time in advertising – it is not that important what people actually say about the product, as long as they talk about it.

Zajonc (1968) goes one step further than just stating the fact that people prefer known things to the unknown. He presents experimental results that allow for the explanation of the effect. The first part of the article describes correlation studies with the use of frequency ranking of English words (Thorndike and Lorge 1944). Zajonc observes that more positive words are more frequently used which may lead to two possible conclusions: either people use more words with positive meaning or the words that are used more frequently become more positive in the process. To check the latter hypothesis he proposed three experimental studies that established the canon of mere exposure paradigm methodology for the following years. The experiments were divided into two parts. In the first part the subjects were presented with nonsense words (experiment 1), Chinese ideograms (experiment 2) or unknown faces (experiment 3) in such a way that each stimulus could appear from 1 to 25 times. In the second part the stimuli were presented once more, together with the stimuli that had never been presented before. After each presentation the subjects were asked to evaluate the positivity of each word, sign or face. The results were compared to the control group who only participated in the second part of the study.

In the following years the mere exposure phenomenon was studied in hundreds of experiments with the use of all sorts of different stimuli: Chinese ideograms, schematic drawings and pictures, faces, possible and impossible 3D shapes and melodies; alterations to the original methodology have been made: forced choice or free evaluations of preference, liking or mood, behavioral and physiological indicators. Detailed descriptions of the experiments, together with the results, can be found in the meta-analysis made by Bornstein (1989), covering 206 experiments. In most of the cases the results confirm Zajonc's hypothesis that people unconsciously prefer things that are known. Additionally Bornstein specifies five characteristic features of the mere exposure effect:

• The effect is stronger when the presentation time is very short or a stimulus is presented subliminally.
• After a relatively low number of stimulus' presentations (10–20) the affective reaction ceases to increase.
• A longer interval (even up to 2 weeks) between presentation and the evaluation leads to a stronger effect.
• The actual recognition is not necessary (and sometimes even disturbing) for the effect to occur.
• Age turns out to be a very important factor modifying the effect in such a way that it is not observed (or even is reversed) in children.

There are several models explaining the mere exposure effect. First theories, dating back to the time before Zajonc referred to the notion of familiarity, come from the tradition of William James (1890). In these approaches it was stated that the known stimuli produce in the mind a warm feeling of intimacy. In the light of the new discoveries, especially the possibility to induce the effect with subliminal presentations, this hypothesis could no longer be held. However the general line of explanation cannot be totally rejected. Similar insights can be found in the processing fluency approach, which we describe in detail below.

After the publication of Zajonc's paper three major explanations, based on different theoretical traditions, were proposed: a two-step model (Berlyne 1970), an optimal arousal model (Berlyne 1974) and an opponent process theory (Harrison 1977).

In the two-step model the mere exposure effect is explained by two interlocked processes: habituation and boredom. Habituation is connected with the feeling of security and therefore manifests itself in positive affect. Excessive exposure leads to satisfaction of curiosity caused by the stimulus which subsequently leads to boredom. The curve portraying the relation between the number of presentations and the positive affect has a shape of an inverted letter 'u', where the initial fast growth is caused by gradual decrease of anxiety caused by an unknown stimulus. The middle plateau relates to the situation when the stimulus is known but does not cause boredom yet and the final decline is a result of growing boredom.

In a similar vein the phenomenon is explained by the optimal arousal theory. Here also authors concentrate on the characteristic shape of the curve. According to Berlyne (1971) one of the basic motivational mechanisms that plays a crucial role in the mere exposure effect is the organism's striving to maintain the optimal level of arousal. Deviation from this level is unpleasant and manifests itself with negative affect. In mere exposure, too few presentations is below optimal level whereas too many lead to over-stimulation.

Also the opponent process theory is based on motivational mechanisms. According to this approach every stimulus causes affective reaction and when it is removed the reaction is reversed. Originally the theory was used to explain sudden shift of mood in the situation of removal of the stressing stimulus. When, after a long stress caused by a fear of having a tumor, a person learns that he is not ill, the mood changes directly from depression to elation. In the case of mere exposure Harrison (1977) postulated that an unknown stimulus causes negative reaction that turns positive when it becomes known – so the source of the anxiety is removed.

All three models fit very well to many of the experimental results, yet all share one major shortcoming. They cannot account for the subliminal mere exposure effect. The fact that with very short presentation times the effect gets even stronger indicates that the process happens originally on a very low, subliminal level and the conscious reaction is but an epiphenomenon. The theories presented above were not originally created with the mere exposure effect in mind; they are just applications of existing approaches. As such they rather try to fit the effect into the frame of an already known mechanism.

In contrast to previous approaches, Moreland and Zajonc (1977) proposed a mechanism dedicated to the effect itself. They have concentrated on the relation of cognitive and affective elements in its creation. According to the researchers there are two possibilities: either there is one common factor responsible for both affective reaction and recognition or there are two distinct factors – subjective affect and subjective recognition – that are later interpreted by the consciousness in accordance with the nature of the task. The analysis of the structural equations model conducted on the results of the experiment confirmed the second hypothesis. Fast affective response turned out to be independent from slower cognitive elaboration of the stimulus. The difference of speed is a result of two different ways of transmitting signals in the brain. The so called lower way used by affective signals is shorter, whereas the upper way which involves the cortex is longer. That is the reason why affect can precede recognition or can appear without conscious reaction. As Zajonc (1980) wrote: preferences need no inferences – we do not have to think to know if we like a given stimulus or not.

The lower affective way enables classification of stimuli that cannot be consciously distinguished. It was shown by Kunst-Wilson and Zajonc (1980) in a study very often cited later by other authors. In the experiment the subjects were presented with random polygon shapes, each shown for 1 ms. During the evaluation phase two stimuli were presented together – one new and one presented before. In a forced choice procedure participants had to choose the stimuli they liked more or decide which of the stimuli was new. The affective reaction turned out to be a better predictor of novelty (60 %), i.e. the stimuli that were presented before were liked more than new stimuli. Recognition remained on chance level. The result was replicated many times (Bornstein et al. 1987; Seamon et al. 1984) so it seems reliable.

The interpretation proposed by the authors focuses on the primacy of affective reaction and its independence from more 'cognitive' recognition processes. However it occurred that with slightly longer presentation times (2–3 ms) the accuracy of recognition went up to the level comparable with the affective reaction. Is it really a strong proof of independence of the processes? This question was raised from the very beginning. Just after the publication of Moreland and Zajonc's paper, Birnbaum and Mellers (1979) pointed out its methodological shortcomings. They argued that based on the available data it was not possible to reject the hypothesis about the single mediator of mere exposure effect. Debate between the researchers lasted for some time and finally the 'two factors' approach won because the two distinct ways of information transfer was consistent with neurobiological data (LeDoux 2000). Yet one might ask: how is it possible that the affective signals 'know' if a stimulus is known or not? Zajonc replies that the stimulus is compared at a very low neuronal level to the patterns stored in the long term memory. While this is probably the case, it does not explain the nature of the mechanism itself. And what is the purpose of the 'cognitive' upper way when the stimulus has been already recognized on a low level? Further in this chapter models based on the dynamics of neural networks will be presented that can account for this strange and paradoxical phenomena. Before, it is necessary to take a closer look at the theories of general meta-cognitive mechanisms of novelty recognition, of which the mere exposure effect is just one.

Meta-Cognitive Mechanisms of Novelty Recognition

It was as early as 1890 when in the first volume of The Principles of Psychology William James described the phenomenon known as the tip-of-the-tongue (TOT) effect. It refers to the situation when we are aware of possessing a piece of information which we cannot remember at the very moment. Here are James' observations:

> Suppose we try to recall a forgotten name. The state of our consciousness is peculiar. There is a gap therein, but no mere gap. It is a gap that is intensely active. A sort of wraith of the name is in it, beckoning us in a given direction, making us at moments tingle with the sense of our closeness, and then letting us sink back without the longed-for term. If wrong names are proposed to us, this singularly definite gap acts immediately so as to negate them. They do not fit into its mould. And the gap of one word does not feel like the gap of another, all empty of content as both might seem necessarily to be when described as gaps. (James 1890)

This short citation contains most of the issues that will reappear in later theories. The most important being the metaphor of a memory trace as a gap. On the one hand the borders of a gap are very precisely defined, so that nothing, except for the memory one is looking for, is going to fit. At the same time a gap is but the border or a shape without content, so it cannot enter the consciousness because it has neither meaning nor a name. The natural conclusion would be that the fact of having a piece of information stored in the memory is not equivalent to being aware of what it is. These are two distinct phenomena.

In contemporary experimental psychology the general version of the effect described by James is known by the name of feeling-of-knowing (FOK) and has been studied from the 1960s. In a classical experiment by Hart (1965) the experimental paradigm called 'recall-judgment-recognition' (RJR) was used for the first time and from then on has been a standard procedure for testing the FOK. It is a very precise translation of James' insights into the language of experimental research. The procedure is divided into three parts. First the participants answer a series of questions concerning general knowledge (e.g. geography or culture). Then a set of questions for which the participants have given wrong answers is separated out. For every question from this set the participants rate the chance that they could give a correct answer if they would be choosing from a set of a few possibilities. Finally, in the last part subjects are doing exactly this: they pick one answer from a set of possibilities prepared by the researcher. The main result of the procedure is the comparison between the subjective judgment and the actual performance. It is an indicator of the strength of the feeling-of-knowing (FOK). Using this paradigm a number of experiments have been conducted with different kinds of stimuli, including names of places, celebrities, TV presenters, acquaintances etc.; different populations have been tested: students, children, elderly people, mentally ill etc. (an extensive review can be found in Nelson et al. 1984). Typically the results show that people judge their unrecallable knowledge above the chance level, although far from 100 %. It suggests that there exists a special mechanism, able to access information in the memory but that cannot be retrieved.

There are two major kinds of explanations of the FOK phenomenon. The first is based on the assumption that people potentially have access to all the information stored in memory and the inability to retrieve it is caused by different kinds of noise interfering with the process. In the theory proposed by Hart (1965) the stimulus has to be strong enough to exceed two kinds of thresholds to be recognized and recalled. First is the FOK threshold (F) and second the recall threshold (R). When the strength of the signal is above R, the stimulus is recognized as familiar, but if it stays in the $<F,R>$ range, a FOK phenomenon occurs.

In the cognitive approaches (Koriat and Lieblich 1974) more stress is put on the way information is stored in memory. Each object is encoded as a referent (a concept) and by its label (a name). The inability to fully recall a memory is due to different interrupts in the system: either an object is properly linked to its reference but the label is missing or, the other way around, the label is recalled but it is not connected with a referent. In the first case it is a TOT kind of experience, in the second case the object can be named but it cannot be understood.

In the approaches of a second kind of explanation of the FOK, it is assumed that judgments of knowledge are based on information that is not directly connected to the stimulus itself. The judgment of knowing is based more on the processing of a question than trying to retrieve the answer. The simplest example is the availability heuristic (Tversky and Kahneman 1974). Former experience with a context similar to the task situation leads to a feeling of familiarity with the stimuli and, based on this feeling, a person judges their knowledge. The inaccuracy of judgment is caused, not by the noise as in previous approaches, but rather by inadequate perception of the situation and oneself. It is clearly visible in cases of FOK induced by social stimuli (Koriat and Lieblich 1974) when judgment is based on the perceived difficulty of the task or a conviction that one has to know a given thing to be considered intelligent. Here FOK is not a cognitive mechanism but is consciously used for self-presentation.

Lately Asher Koriat (2000) described FOK as a meta-cognitive mechanism that is a bridge between two levels of information processing. On the higher level, information is processed under the control of the consciousness; on the contrary the lower level is automatic and unconscious. There are judgments based on either of the levels – analytic or automatic. However there are processes, of which a TOT phenomenon is a good example, that require passing information between the two levels. In the tip-of-the-tongue experience an unconscious process of information retrieval invokes in the consciousness a feeling of familiarity that in turn leads to engagement of higher level heuristics for the searching process. The result of the search is continuously monitored by the lower level process that feeds back the affective information.

Fluency

Processing fluency is a quick and efficient reaction to a stimulus, however this general definition should be made more precise. Information processing can be described on various levels not related directly to the stimulus. Mental representations can vary in levels of arousal (Mandler 1980), stimuli can be processed at different speeds (Jacoby 1983) and commitment (Schwarz 1998). In spite of their obvious dissimilarities, all those dimensions are often referred to as fluency. Many researchers agree that the fluency signals coming from different sources influence the information processing by way of a meta cognitive feedback mechanism (Mazzoni and Nelson 1995; Metcalfe and Shimamura 1994) often connected with the affective system (Fernandez-Duque et al. 2000).

Fluency may or may not manifest itself in consciousness. We can, therefore, make a distinction between objective fluency related to fast and efficient processing that does not use much cognitive resource and a subjective feeling of fluency that often leads to a positive affective reaction and the impression of easiness and efficiency. Both processes can be, to some extent, independent. Well studied or automatic actions that are performed with a high level of objective fluency can pass unnoticed by the subject, just because it is so fluent. On the other hand, a person can experience a very high level of subjective fluency while performing very poorly. The best example occurs in the state of being drunk (Winkielman et al. 2002).

Processing fluency is a global phenomenon which can come from different cognitive levels and different structures pointing to various cognitive processes. We can further distinguish between perception fluency, referring to a perceived object's external features (Bornstein and D'Agostino 1994; Jacoby 1983; Reber and Schwarz 1999; Reber et al. 2004; Winkielman et al. 2002) and semantic fluency, referring to structural or abstract features of processed stimuli (Kelley and Jacoby 1998; Whittlesea 1993; Winkielman et al. 2002). Each type is prone to a different kind of manipulation. Perceptive fluency can be caused by simple repetitions, priming with shapes and colors and also varying contrast or presentation time. Semantic fluency is triggered by semantic priming, coherence with context and formal features such as rhythm or rhyme etc. Both types of fluency manifest themselves differently in the brain. The perceptive fluency effect is visible mainly in sensory cortex, whereas semantic fluency is responsible for the reactions in prefrontal cortex.

As was mentioned above, subjective fluency often manifests itself in the consciousness as positive affect or so called cognitive feeling, that is intuition or a presentiment etc. The affective response can be, and often is, transferred to the stimulus. This effect vanishes when a person is provided with different possible attributions for the experienced state (Schwarz and Clore 1996; Winkielman et al. 2002). In a study conducted by Fazendeiro et al. (2005), the semantic priming procedure was used. Subjects judged presented words, of which some were primed by other words either semantically close or far from them. Additionally, subjects heard music playing in the background. One group was told that music can improve the performance speed. In this group there was no difference with the control group, i.e. primed words were judged as more positive. In the other group it was suggested that music can be the real cause of the emotions felt during the study. This time the effect of affective feedback was gone and all stimuli were judged independently from the type of priming. In means that the spontaneous interpretation of subjective affective signals is tied to the assessment of currently processed stimulus. Once the alternative source for attribution is provided, the effect vanishes.

This very feature caused the effect of fluency to be used as a possible explanation of the mere exposure effect (Bornstein and D'Agostino 1994; Jacoby et al. 1989; Seamon et al. 1983a, b; Whittlesea 1993). The positive reaction towards neutral stimuli indicates a similar mechanism behind both phenomena. Should the feeling of fluency be responsible for the affective reactions, the results characteristic of the mere exposure effect would be obtained manipulating processing fluency. In one of the studies conducted with the aim of verifying this hypothesis, subjects were presented with a set of pictures of everyday objects. The quality of the pictures was diminished so that the task of recognizing them was non trivial. Some of the pictures were subliminally primed with either matching or mismatching contours. Priming served as the high fluency manipulation; indeed, subjects preferred the objects presented in the high fluency condition. The stimuli did not differ in the level of familiarity (the mere exposure condition) so the only cause of the effect analogous to mere exposure could have been the fluency. Similar results were obtained using presentation time and figure-ground contrast manipulation (Reber et al. 2004).

Similarly to explanations of the mere exposure effect, processing fluency was often explained using the two-step models referring to the logic of Schachter and Singer's (1962) theory of emotions. The first step of emotion creation is affectively neutral arousal which is subsequently interpreted by the cognitive system as a positive or a negative emotion, based on the available contextual cues. According to this logic the fluency signal would initially be neutral and would obtain its valence from the context. Mandler, Nakamura and Van Zandt (1987) proposed to interpret the primary arousal as the increased cognitive availability of the stimulus. This in turn leads to formation of more polarized judgments, the valence of which depends on the external circumstances. If we ask subjects to choose the objects of the lightest color, then the high fluency would translate to the color lightness; but if we asked them to find the darkest ones, the situation would be reversed and the same fluency level would be translated to the color darkness. A slightly different approach can be found in Jacoby, Kelley and Dywan (1989). They treat fluency as a sort of affectively neutral arousal that can be interpreted according to the situation as familiarity in the mnemonic task or a feeling of confidence in a problem-solving task. Finally, Bornstein and D'Agostino (1992) propose to treat the mere exposure effect as nothing more than trying to find a simplest explanation for the experienced (neutral) feeling of fluency.

Two-step models of fluency, such as the Schacter and Singer's theory of emotions, do not always find clear empirical support. As far as the judgments of lightness or similarity are concerned the results follow the assumptions. However, different kinds of stimuli often lead to asymmetrical responses, where fluently processed positive stimuli are more liked but fluently processed negative stimuli do not change the judgment (Seamon et al. 1998). A very sharp and clean picture of an ugly object does not make it more ugly but neither does it not make it less ugly. The supporters of two step models say that a 'liking' judgment is more natural than 'disliking'. Highly fluent processing may easily be interpreted as 'liking' but 'disliking' requires more complicated mental operations which accounts for the results. Winkielman et al. (2002) observe that the asymmetry is characteristic of evaluative judgments, such as 'like–dislike', and is not present in more complicated tasks. They say that fluency signals are processed in the very first moments of stimulus recognition when other cues are not yet available. Therefore, they indicate nothing more than the general easiness of information processing. After the meaning is attached to the object, the fluency-based response sums up with its perceived valence. In the case of positive stimuli, the overall response is the increased or decreased 'liking'. The negative stimuli lead to a more complicated scenario: more fluently processed negative stimulus is better accessible to consciousness and can more easily be rated as negative, this judgment however is balanced by the positive signals from the ease of fluency itself. On the other hand, a when a stimulus triggers low fluency signals, it means it is not so well perceived and its negativity is less, but at the same time this negativity is increased by the low fluency of the signals. The asymmetry comes then from the reinforcement of positive and canceling of negative judgments.

To further support the hypothesis that fluency signals act at the early stages of signal processing against the two-step theories, Winkielman and Cacioppo (2001) conducted an experiment using an EMG (electromyography) procedure. They replicated the study described earlier in this section, where subjects had to evaluate pictures of objects from everyday life, primed by either coherent or incoherent shapes. Before participants' actual responses the reaction of muscles in the face responsible for smiling and frowning were measured. The results show a high correlation between processing fluency and the 'smiling' muscle activity. This proves that the fluency signals are processed prior to higher cognitive operations such as, for example, attribution.

The role of processing fluency as well as other meta-cognitive signals described above, such as FOK, is to provide feedback information on the course of different cognitive processes. For this reason such signals have to operate fast and independently from higher level mechanisms. They have to act in parallel with the process they are monitoring. It leads to a paradoxical situation where one is able to assess the stimulus before it has been recognized. What is more, the recognition process can be altered by certain features of a stimulus even before it has actually been noticed. This paradox can be resolved only with the model of a mechanism able to monitor its own dynamics. It would not have to be triggered by any external structures so no other structure would have to be aware of its existence and it could spontaneously produce meta-cognitive signals. Such a model was proposed by Lewenstein and Nowak (1989). It was based on an artificial neural network, so before describing its details it will be useful to have quick glimpse on the basics of artificial neural network theory.

Artificial Neural Networks

An artificial neural network (ANN) is a set of simple elements, often called neurons, connected with links, often called synapses. The neurons in an ANN resemble the biological cells only on a very general level. According to the classic

McCulloch and Pitts' model (1943), each neuron is a simple processor that consists of input – receiving signals from other neurons, an activation function – defining its response to incoming signals, and output – transmitting signals to other neurons. There exist of course more complicated and biologically accurate models but this one is widely used in computer simulations, including those that are going to be presented in this chapter. In this simplified approach, the whole process taking place in a cell is reduced to binary signal processing. Each neuron receives information from multiple other neurons connected to it, then the signal is summed up with the use of a certain activation function and, at the end, the output is produced in a form of '0' or '1'. The activation function may vary but it has to be nonlinear, with a threshold. Commonly used functions are: logistic function, hyperbolic tangent or Gauss curve.

The binary neurons can hold only a limited amount of information, but the real heart of a network is the structure of connections. Each synapse connects two neurons and transmits information about their state, modified by a continuous parameter called weight The activity of most networks can be divided into two distinct phases: learning and recognition. In the learning phase the weights of synapses are modified according to a learning rule. In the recognition phase the network is activated with a stimulus or a set of stimuli and generates a response based on the connection matrix. The response is a stable pattern of activations of neurons that can be associated with a certain meaning.

There are at least two general types of ANN: the feed-forward multilevel networks and the attractor networks. The first kind will not be used in the models described in this chapter so only a short summary of their characteristics will be given. They were invented in 1958 by Rosenblatt as a perceptron – a simple system implementing logical operations such as AND and OR. The model has been extended many times by many authors and finally came to a form of multilayered structure where each neuron is connected with all neurons in its adjacent layer. Information is fed to the network in the form of an activation vector consisting of the states of every neuron in the entry layer. The vector is then transmitted to other layers, changing according to the weights of connections. At the end the output vector is produced. The most popular learning algorithm used in feed-forward networks is back-propagation, where the weights of all neurons are changed globally in such a way that the output vector matches the desired networks response.

The other class of models are so called attractor neural networks. This architecture was described by Hopfield (1982) as a model of stochastic phenomena in physics. In Hopfield's network all neurons form a single layer and are connected to each other with symmetrical connections. In the learning phase the so called Hebb rule is used. In this algorithm a synapses weight is proportional to the correlation of the activities of neurons on both its ends. If the two neurons happen to be in the same state at a given moment, the weight goes up; otherwise it decreases. Remembering a pattern consists of a one-time modification of weights. Each new pattern is added to those already known. If in the recognition phase a known pattern should be presented, no network reaction is going to be observed. However, if a

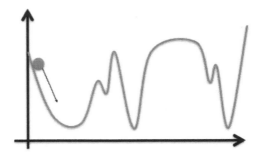

Fig. 4.1 The energy landscape. The *line* represents values of energy function for different system states. The *ball* represents the actual state of the system and the vector attached to it, the direction and magnitude of change

modified or damaged pattern is presented the network will try to fit it to the closest one stored in the memory. The memory is not unlimited and after exceeding a certain level (14 % of the number of neurons) the recognition becomes less and less precise.

No matter if the network has anything stored in memory it always converges on a stable state. This effect is related to the so called energy function. The energy in a Hopfield network can be defined as the distance of current system's state from the stable state. Using the notion of energy, the space of possible states of the network can be portrayed as the energy landscape – a function that assigns the energy level to each possible vector of states. The valleys in the plot of the energy function signify the areas of low energy and in most cases are related to the remembered patterns. If the system is initialized in one of such low energy places it will quickly stabilize there and no dynamics will be observed. However if it should be initialized in one of the high energy peaks the network will work until the state reaches the lowest possible level of energy. The areas of lowest energy are called the attractors as they pull the system towards themselves.

The shape of the energy landscape can be treated as the systems memory. On the schematic picture (Fig. 4.1) it is represented as a curve on a plane; in fact it is a multidimensional surface of the dimensionality N equal to the number of neurons. The momentary state of the system is represented on the picture by the small ball. If in the immediate surroundings of the ball a point of lower energy is found, the circle moves towards it with the speed proportional to the difference of energy levels. Once the 'valley' is reached, the ball stays there until an external force is applied. The special kind of external influence, called 'noise', is used to ensure that the system does not end up in local minima but tries to find a real minimum with the lowest energy. The noise can be implemented as a mechanism randomly switching the states of a certain number of neurons. It can be compared to 'shaking' the energy landscape strongly enough for the ball to fall out of shallow shelves or niches but not out of the deep valleys.

Simulation Models Based on Coherence Signals

The energy in a Hopfield neural network is a direct result of reciprocal interactions of neurons. Lower energy can generally be connected with lower network activity, i.e. a smaller number of changes in individual neurons' states. Neurons change their state if this is not in accord with the summary signal coming from other neurons. It can be said that the energy is related to the average fit of all the neurons to the global network activation level. This fit is often called coherence and it is a global indicator of network performance.

High coherence is characteristic of processing well-known stimuli and, analogically, low coherence signifies of novel stimuli. On the basis of the coherence level connected with the process of recognition it is possible to distinguish the known stimuli from the novel even before they have been fully recognized. When the ball starts to move down the steep slope towards a valley, even though the movement has only just started, with high probability it can be predicted that the ball is going to reach the bottom and stay there. Similarly in the process of recognition, if a rapid decrease of energy level is observed, it can be inferred that the system has started on the path of fully recognizing the stimulus and therefore even though the stimulus has not been identified yet, it is probably known to the system.

There are several indicators of the current level of coherence. The simplest one is 'volatility' and it is defined by the number of neurons changing state in a given moment. When a network is approaching the attractor more and more neurons are in their final states and do not change any more. Once the minimum is reached the volatility index drops to zero. Analogically, when the system is far from any attractor, the states of most neurons are not stable and the volatility index stays high. As it was said before, the attractor network always reaches a stable point, even when it was activated by an unknown stimulus. So how should it be possible to tell known stimuli from the new ones? The temporal pattern of volatility index is different in either situation. In the case of known stimuli its value drops rapidly and stays low until the end of the recognition process. When novel stimuli are processed the volatility stays high for a longer time before dropping to zero. This difference in the dynamics of the first moments of the recognition process allows for the fast assessment of novelty.

Other coherence indicators include average local field level and signal-to-noise ratio. The first is based on the coherence of the signal coming to a neuron from other neurons. It is higher in the vicinity of an attractor. The latter is a comparison between average noise signal and the average neuron's activity signal. Again it is higher close to attractor, where the neurons act in synchrony and low in the other regions where neuron activity is close to random. Both indicators show similar dynamics and the gap between known and novel stimuli can be observed.

The novelty recognition mechanism has been applied in several simulation models. Lewenstein and Nowak (1989) showed a system that regulated itself by increasing the noise level in low coherence states. It prevents the system from reaching the local minima and the energy from dropping to zero when processing unknown or ambiguous stimuli. A further step is the model called 'Smartnet' (Żochowski et al. 1993) where coherence signals were used as a moderator of global system behavior. When a stimulus was judged known, the recognition process would go on, but novel stimuli caused the system to switch to a learning phase. Finally, a model proposed by Drogosz and Nowak (2006) recreated the mere exposure effect with the use of coherence level monitoring. Just as in classic experiments, the response of the system in very short presentations was based solely on nonspecific feedback information – in this case coherence monitoring.

Coherence Monitoring as the Model for Metacognitive Mechanism for Novelty Recognition

Intuition suggests that in order to take action towards an object or simply evaluate, it is necessary to notice and recognize it. In such a common-sense model the processing of a stimulus would start with the classification of its features followed by classification to one of the remembered categories. Next, it would be evaluated and the appropriate behavioral reaction could be taken. The analysis of empirical results obtained in various paradigms leads to rather different conclusions. In most of the cases, the reaction to a stimulus either precedes or is independent from conscious recognition. This leads to an apparent paradox, since it seems impossible to assess an object before it is even identified. If we look more closely at the assumption that nothing can be said about a stimulus until after it is recognized, we are going to realize that this is a kind of circular argument. For if a complete recognition is at all possible, it means that a stimulus in question was a known one. Should it be unknown, the recognition process would not have the right to come to an end at all. This leads to a conclusion that if we actually observe such phenomena, they have to be based on the recognition process itself and not on its results.

Novelty is not an objective feature of an object, it is rather an individual reaction of a given subject perceiving it. Exactly the same thing may be well-known to one person and be totally new to another. It is the cognitive system processing a stimulus that is the source of the novelty classification. The explanations of phenomena, such as mere exposure or feeling-of-knowing, take this into account by introducing a sort of affective or cognitive feedback mechanism that is supposed to operate in parallel with the recognition process. However, they do not provide a simple mechanism modeling such a system. The coherence monitoring seems a perfect solution. It is built in to any cognitive system operating on the basis of a neural network model. No additional monitors or probes have to be added. It doesn't necessarily mean that all those phenomena are implemented with the same attractor neural network architecture as the one described above. The mechanisms working in the brain are certainly more diversified and complex, however they all share the distributed nature of processing where notions such as energy or coherence can be defined. Once it is available, all sorts of feedback loops can be implemented.

In computer simulations, the feedback mechanism is implemented as a part of the algorithm, simply calculating the dynamics of neurons and using the result as the system's response or as its moderator. In a real person the information has to manifest itself in a form of a phenomenon accessible to consciousness. This is exactly what the meta-cognitive mechanisms are. In this chapter several effects were described including mere exposure and feeling-of-knowing. All of those mechanisms rely on meta-cognitive feedback. In the case of mere exposure it is the increased preference for known stimuli; in FOK, or tip-of-the-tongue, it is a kind of intuition or cognitive feeling, the objective processing fluency is marked by positive affect. More examples could be given, such as all sorts of implicit learning phenomena, cognitive monitoring as well as cognitive illusions of the kind of déjà vu. In each case the judgment or evaluation is very fast and in most cases precedes the awareness. Therefore to each of those situations the logic used above applies: if it is faster than conscious realization, then the result cannot be based on this realization but has to be inferred from the process.

The point of this chapter was to show how the feedback mechanism postulated by many classic theories could be easily and parsimoniously implemented with use of the notion of coherence. It was shown as a generic theory rather than a ready-made model of any particular process. It can be adapted to specific processes by identifying necessary elements such as distributed memory representations and by defining coherence indicators such as the energy function.

References

Berlyne, D.: Novelty, complexity, and hedonic value. Percept. Psychophys. **8**, 279–286 (1970)

Berlyne, D.E.: Aesthetics and Psychobiology. Appleton, New York (1971)

Berlyne, D.E. (ed.): Studies in the New Experimental Aesthetics: Steps Toward an Objective Psychology of Aesthetic Appreciation. Hemisphere, Washington (1974)

Birnbaum, M.H., Mellers, B.A.: Stimulus recognition may mediate exposure effects. J. Pers. Soc. Psychol. **37**(3), 391–394 (1979)

Bornstein, R.F.: Exposure and affect: overview and meta-analysis of research, 1968–1987. Psychol. Bull. **106**, 265–289 (1989)

Bornstein, R.F., D'Agostino, P.R.: Stimulus recognition and the mere exposure effect. J. Pers. Soc. Psychol. **63**, 545–552 (1992)

Bornstein, R.F., D'Agostino, P.R.: The attribution and discounting of perceptual fluency: Preliminary TESTS of a perceptual fluency/attributional model of the mere exposure effect. Soc. Cogn. **12**, 103–128 (1994)

Bornstein, R.F., Leone, D.R., Galley, D.J.: The generalizability of subliminal mere exposure effects: influence of stimuli perceived without awareness on social behavior. J. Pers. Soc. Psychol. **53**, 1070–1079 (1987)

Drogosz, M., Nowak, A.: A neural model of mere exposure: The EXAC mechanism. Polish Psychol. Bull. **37**(1) 7–15 (2006)

Fazendeiro, T., Winkielman, P., Luo, C., Lorah, C.: False recognition across meaning, language, and stimulus format: conceptual relatedness and the feeling of familiarity. Mem. Cognit. **33**, 249–260 (2005)

Fechner, G.T.: Vorschule der Ästhetik. Breitkopf & Härtel, Leipzig (1876)

Fernandez-Duque, D., Baird, J.A., Posner, M.I.: Executive attention and metacognitive regulation. Conscious. Cogn. **9**, 288–307 (2000)

Festinger, L.: Informal social communication. Psychol. Rev. **57**, 271–282 (1950)

Harrison, A.A.: Mere exposure. In: Berkowitz, L. (ed.) Advances in Experimental Social Psychology, vol. 10, pp. 39–83. Academic, San Diego, CA (1977)

Hart, J.T.: Memory and the feeling-of-knowing experience. J. Educ. Psychol. **56**(4), 208–216 (1965)

Hopfield, J.J.: Neural networks and physical systems with emergent collective computational capabilities. Proc. Natl. Acad. Sci. **79**(8), 2554–2558 (1982)

Jacoby, L.L.: Perceptual enhancements: persistent effects of an experience. J. Exp. Psychol. Learn. Mem. Cogn. **9**, 21–38 (1983)

Jacoby, L.L., Kelley, C.M., Dywan, J.: Memory attributions. In: Roediger, H.L., Craik, F.I.M. (eds.) Varieties of Memory and Consciousness. Essays in Honor of Endel Tulving, pp. 391–422. Lawrence Erlbaum, Hillsdale, NJ (1989)

James, W.: The Principles of Psychology, vol. 1, Dover Publications, New York (1950). (Original work published 1890)

Kelley, C.M., Jacoby, L.L.: Subjective reports and process dissociation: fluency, knowing and feeling. Acta Psychol. **98**, 127–140 (1998)

Koriat, A.: The feeling of knowing: some metatheoretical implications for consciousness and control. Conscious. Cogn. **9**(149), 171 (2000)

Koriat, A., Lieblich, I.: What does a person in a "TOT" state know that a person in a "don't know" state doesn't know? Mem. Cognit. **2**, 647–655 (1974)

Knust-Wilson, W.R., Zajonc, R.B.: Affective discrimination of stimuli that cannot be recognized. Science **207**, 557–558 (1980)

LeDoux, J.E.: Emotion circuits in the brain. Annu. Rev. Neurosci. **23**, 155–184 (2000)

Lewenstein, M., Nowak, A.: Recognition with self-control in neural networks. Phys. Rev. A **40**(8), 4652–4664 (1989)

Mandler, G.: Recognizing: the judgment of prior occurrence. Psychol. Rev. **87**, 252–271 (1980)

Mandler, G., Nakamura, Y., Van Zandt, B.J.: Nonspecific effects of exposure on stimuli that cannot be recognized. J. Exp. Psychol. Learn. Mem. Cogn. **13**, 646–648 (1987)

Mazzoni, G., Nelson, T.O.: Judgments of learning are affected by the kind of encoding in ways that cannot be attributed to the level of recall. J. Exp. Psychol. Learn. Mem. Cogn. **21**, 1263–1274 (1995)

McCulloch, W.S., Pitts, W.: A logical calculus of the ideas immanent in neural nets. Bull. Math. Biol. **5**(8), 115–133 (1943)

Metcalfe, J., Shimamura, A.P.: Metacognition: Knowing about Knowing. MIT Press, Cambridge, MA (1994)

Moreland, R.L., Zajonc, R.B.: Is stimulus recognition a necessary condition for the occurrence of exposure effects? J. Comp. Physiol. Psychol. **35**, 191–199 (1977)

Nelson, T.O., Gerler, D., Narens, L.: Accuracy of feeling-of-knowing judgments for predicting perceptual identification and relearning. J. Exp. Psychol. Gen. **113**, 282–300 (1984)

Reber, R., Schwarz, N.: Effects of perceptual fluency on judgments of truth. Conscious. Cogn. **8**, 338–342 (1999)

Reber, R., Schwarz, N., Winkielman, P.: Processing fluency and aesthetic pleasure: Is beauty in the perceiver's processing experience? Pers. Soc. Psychol. Rev. **8**, 364–382 (2004)

Schachter, S.E., Singer, J.: Cognitive, social and physiological determinants of emotional state. Psychol. Rev. **69**(5), 379–399 (1962)

Schwarz, N.: Accessible content and accessibility experiences: the interplay of declarative and experiential information in judgment. Pers. Soc. Psychol. Rev. **2**, 87–99 (1998)

Schwarz, N., Clore, G.L.: Feelings and phenomenal experience. In: Higgins, E.T., Kruglanski, A.W. (eds.) Social Psychology: Handbook of Basic Principles, pp. 433–465. Guilford, New York (1996)

Seamon, J.G., Brody, N., Kauff, D.M.: Affective discrimination of stimuli that are not recognized: effects of shadowing, masking, and cerebral laterality. J. Exp. Psychol. Learn. Mem. Cogn. **9**(3), 544–555 (1983a)

Seamon, J.G., Brody, N., Kauff, D.M.: Affective discrimination of stimuli that are not recognized: II. Effect of delay between study and test. Bull. Psychon. Soc. **21**(3), 187–189 (1983b)

Seamon, J.G., Marsh, R.L., Brody, N.: Critical importance of exposure duration for affective discrimination of stimuli that are not recognized. J. Exp. Psychol. Learn. Mem. Cogn. **10**, 465–469 (1984)

Seamon, J.G., McKenna, P.A., Binder, N.: The mere exposure effect is differentially sensitive to different judgment tasks. Conscious. Cogn. **7**, 85–102 (1998)

Sokolov, E.N.: Perception and the Conditioned Reflex. Pergamon, Oxford (1963)

Thorndike, E.L., Lorge, I.: The Teacher's Word Book of 30,000 Words. Teachers College, Columbia University, New York (1944)

Tversky, A., Kahneman, D.: Judgement under uncertainty: heuristics and biases. Science **185**, 1124–1130 (1974)

Whittlesea, B.W.A.: Illusions of familiarity. J. Exp. Psychol. Learn. Mem. Cogn. **19**, 1235–1253 (1993)

Winkielman, P., Cacioppo, J.T.: Mind at ease puts a smile on the face: psychophysiological evidence that processing facilitation leads to positive affect. J. Pers. Soc. Psychol. **81**, 989–1000 (2001)

Winkielman, P., Schwarz, N., Fazendeiro, T., Reber, R.: The hedonic marking of processing fluency: implications for evaluative judgment. In: Musch, W.J., Klauer, K.C. (eds.) The Psychology of Evaluation: Affective Processes in Cognition and Emotion, pp. 189–217. Lawrence Erlbaum, Mahwah, NJ (2002)

Zajonc, R.B.: Attitudinal effects of mere exposure. J. Pers. Soc. Psychol. (Monogr. Suppl.) **9**(2), 1–27 (1968)

Zajonc, R.B.: Feeling and thinking: preferences need no inferences. Am. Psychol. **35**, 151–175 (1980)

Zajonc, R.B.: On the primacy of affect. Am. Psychol. **39**(2), 117–123 (1984)

Żochowski, M.R., Nowak, A., Lewenstein, M.: Memory that tentatively forgets. J. Phys. A: Math. Gen. **26**(9), 2099–2112 (1993)

Chapter 5
The Dynamics of Patterns of Commitment in Sports

Dariusz Parzelski and Andrzej Nowak

Abstract Commitment is the major factor in sport; it combines in one variable the separate elements that influence the sporting result that is achieved. Based on sport psychology data and the dynamical Social Psychology theory of Vallacher and Nowak, it was assumed that there is a natural taxonomy relating the degree to which individuals practice sport to their degree of commitment. This research checks the concordance between the type of sport commitment assigned using dynamic measures and those assigned using classic methods of content analysis.

To determine the type of sport commitment both the Mouse Task in a dynamic paradigm and a method using competent judges assessment of participants' statements describing typical daily activity schedules were used. All 67 participants were athletes taking part in sport competitions of different importance. Based on our assumption, four natural types of sport commitment were observed: 'professionalism', 'amateurishness', 'fervor' and 'professional amateurishness'. We show that the computer mouse method captures information concerning this theoretical construct, which is not captured by the classic method of content analysis.

Introduction

For years the interests of researchers have been focused on finding the answer to the following question: what are the common traits that characterize sport champions (Anshel 2003; Williams 1980)? Aside from talent, adequate motivation and motor

D. Parzelski
Sport Psychology Department, Warsaw School of Social Sciences and Humanities, Chodakowska 19/31, 03-815 Warsaw, Poland
e-mail: dparzelski@swps.edu.pl

A. Nowak
Department of Psychology, University of Warsaw, Stawki 5/7, 00183 Warsaw, Poland
e-mail: andrzejn232@gmail.com

A. Nowak et al. (eds.), *Complex Human Dynamics*, Understanding Complex Systems, 67
DOI 10.1007/978-3-642-31436-0_5, © Springer-Verlag Berlin Heidelberg 2013

abilities, a common feature is that, since their early childhood, sport has played a primary role in their lives. Sport influences and organizes an athlete's mode of existence and their proximate surroundings.

The factor which affects the life of an individual and determines the standard that they achieve in sport is the manner of commitment to actions related to sports. Physical activity may play a dominant role in an athlete's life and determine their actions in other fields, such as work, education or family matters. For some athletes sport might be a supplement and setting for major domains of life. A higher level of commitment leads to more extensive functioning in the domain of sports, which results in better achievements. A lower level of sport commitment means weaker contacts with the domain of sports and not such good athletic achievements.

The commitment has an effect on an athlete's sport behavior and might be reflexively shaped by it. While many traditional motivation theories might encompass commitment, they do not capture the crucial aspect of positive feedback, which is characteristic of the phenomenon of sport commitment. Feedback loops capture the dynamic, as opposed to the static character of commitment in action. Note that the level of commitment is not fixed; it alters day by day, or week after week. In an extended time perspective, this level demonstrates specific consistent features, e.g., an athlete, after a period of low intensive work-outs, tends to compensate by increased training, whereas an over-trained athlete strives for relaxation.

Commitment is a structure which combines many factors in one system; between the components of such a structure there occur nonlinear dependencies. To comprehend the nature of such system, a theory capable of explaining the complicated bilateral dependencies is essential. Complex systems theory provides such a perspective as it is based on the assumption that every phenomenon might be conceived as a configuration of mutually linked elements (Vallacher and Nowak 2007). To understand the essence of how sport commitment functions it is crucial to grasp the dynamics between its separate components. This system changes and evolves in time using internally generated dynamics as well as under the influence of external factors. In this way the system tends towards attractors, the stable equilibrium points, which function as the system's regulators.

From a complex systems perspective, sport commitment might be considered to be a dynamic system consisting of psychic mechanisms and the behaviors of an individual, as well as external influences. Using this terminology, it is possible to picture sport's commitment through diverse types of dynamics. These types might differ, inter alia, in the attitude of an individual towards sport and other life domains, the extent to which sport satisfies the athlete's needs and also the amount of time, energy and financial resources they allocate to sport. Particular types may well be connected with higher or lower level of athletic achievements. The concept of sport commitment begins to lose the characteristics of a quantity variable and becomes a quality variable.

Concept of Commitment in Psychology

From a psychological point of view, commitment to a particular action is a basic and intuitively obvious condition for obtaining positive results (Morrow 1993). Commitment fulfills a control function in an individual's life. It determines thoughts, actions and psychological needs of the person in all their life domains.

Lesyk (1998) distinguished athletes' degree of commitment to their sport as follows: 1. The Casual Participant; 2. The Recreational Athlete; 3. The Serious Recreational Athlete; 4. The Committed Athlete; 5. The Elite Athlete. Each consecutive level of commitment parallels an increase in the significance of the role that sport plays in life of an individual. "Generally, as we go up the scale, the person is devoting more time, effort, and resources to the sport and is expecting more in return in terms of performance, achievement, and recognition" (Lesyk 1998, p. 86).

Theoretical Context: Theory of Complex Systems

To understand how heterogeneous psychological and social factors make up the structure of sport commitment, complex systems theory will be employed. From this point of view an individual is treated as a set of many elements connected with one another, each changing in time. The dynamic system is a combination of inter related features, each undergoing change, whose main characteristic is its capability of development over time (Nowak and Vallacher 1998; Vallacher and Nowak 2007).

Using this concept, an athlete's sport functioning can be viewed as a dynamic system which develops in time. The dependence between practice and success is a mechanism of mutually connected feedback loops. Practice leads to success and increased sport commitment but also success, reversibly, affects the amount of practice and its quality – causing the incorporation of new activities e.g., general fitness, strength, mental training. This process can lead to an intensified commitment to actions that are being undertaken. As a result, still more areas of a person's functioning become connected with sport.

Tools of Dynamic Social Psychology

The tools and methods deriving from the complex systems theory have already been used occasionally to study sporting situations, for instance judo fighting (Gernigon et al. 2004). To these tools belong: the mouse procedure, the programs analyzing the patterns of systems' dynamics and computer simulations. These methods have the capacity of analyzing nonlinear dependencies, which occur between many elements in the system of sport commitment of an individual.

One of the crucial tools for examination of system dynamics is a procedure using a computer mouse. The fundamental assumption concerning the application of the mouse procedure is that the variability of the intrinsic dynamic system of an individual may be portrayed either by the observation of a person, or by the analysis of their stream of thought concerning a specific topic. The mouse procedure allows us to make a description of the variation of patterns in the stream of consciousness.

The next tool that belongs to the methods used in complex systems theory, is a program analyzing the patterns of system dynamics (Johnson and Nowak 2002), based on the presumption that in the vicinity of an attractor the system behaves in a stable way. The system's equilibrium point is a state that is frequently reached by a specific configuration and maintains stability for a longer time.

Typology of Sport Commitment

The starting point for dividing the commitment into separate types were studies of Vallacher and Nowak (2007) on 'I' dynamics and studies of Johnson and Nowak (2002) on patterns of dynamics in depressive patients. This research permits us to presume that with the use of the complex systems' paradigm one might describe the various types of commitment that athletes have to their sport. Using the language of complex system theory, the sport commitment model and the continuum of commitment by Lesyk (1998), we distinguish four main types of sport commitment dynamics. Each type is a stable equilibrium point (attractor) of the system which determines the dynamics of an athlete's system and connects with the meaning that the sport has in their life.

'Professional' Type of Commitment

For a person belonging to this group, sport is a dominant activity to which they are ready to resign many other life domains and make many sacrifices in order to achieve success. This is the only way to win in a highly competitive environment. Sport is no longer treated as a means to fulfill one's needs, but rather the fulfillment of needs is brought under the conditions imposed by sport training. The system sets the framework in which the individual's needs may be realized.

The dominant equilibrium point for such an athlete is a punctual (deep, narrow) attractor in the sport domain (Fig. 5.1). Usually the increase of commitment, and additionally the increase in the number of needs fulfilled by sport (deepening and narrowing of the attractor) has a positive influence on the results that are achieved. An athlete shows a high tolerance on such a shift. On the other hand, satisfying some of the needs outside sports has a negative impact on sport commitment and correspondingly on achievements (wider and shallower attractor). The system demonstrates a very low tolerance for such types of change. It forces the system to return to the dominant equilibrium point or 'builds the new attractor' of an individual.

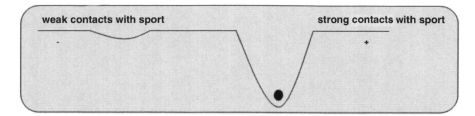

Fig. 5.1 The 'professional' type of sport commitment. The continuum of contacts with sport for an athlete highly committed to sport. The system's equilibrium point finds itself close to the end of continuum describing strong contacts with sport (+). The *ball* symbolizes the dynamic system of an athlete

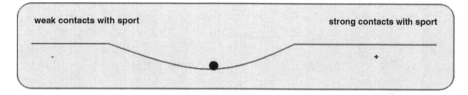

Fig. 5.2 The 'amateur' type of sport commitment. The continuum of contacts with sport for an athlete lowly committed to sport. The system's equilibrium point finds itself close to the end of continuum describing weak contacts with sport (−). The *ball* symbolizes the dynamic system of an athlete

'Amateur' Type of Commitment

Amateurs are people doing sports quite regularly, at a low standard. They train mainly because of their high need for physical activity. Such individuals rarely compete; they train for pleasure. They prefer a specific kind of activity: running, swimming, cycling. Sport is a means of satisfying the need for carrying out the activity itself. It is also a supplement for other life domains that are important for the individual. People belonging to this group frequently do sports because of the influence they are under – friends, family – as this is the way of spending time in their environment. The equilibrium system in this case is shallow and wide, actually just 'touching upon' contacts with sport (Fig. 5.2).

This system is highly sensitive to over-training (exhaustion) and less responsive to under-training. An individual does not resign other activities (work, education, family) in order to take part in practice. The achievement need has no effect on an individual's functioning, because this is satisfied in other life domains. One avoids comparing and competing with others in sport since entering the rigor of external evaluation might destroy the joyfulness of training, and as a result lead to resigning from practice.

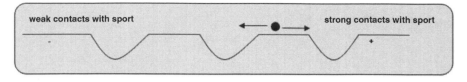

Fig. 5.3 The 'fervor' type of sport commitment. The continuum of contacts with sport for an athlete frequently engaged in sports but not in a permanent way. The system's equilibrium point changes depending on the field in which one fulfills needs. The ball symbolizes the dynamic system of an athlete

'Fervor' Type of Commitment

Athletes belonging to this group do sports at a quite high standard, but still do not achieve successes regularly. They satisfy plenty of needs through sport training. Simultaneously, they are active participants in other life domains, such as: hobby, work, social contacts, family. They gain medals, although rather rarely. Sometimes they are called up to the national team and occasionally represent their country in international competitions. People representing this type of commitment are lacking the dominant equilibrium point. It is rather a system of many shallow and wide attractors (Fig. 5.3).

Sport is a means of satisfying various individual needs of social contact, social approval, physical activity, material needs (journeys, awards), sexual needs, independence, social prestige, self-esteem. Sport is trained until it fulfills these individual needs. At the moment other ways of fulfilling become available, or sport stops gratifying these individual wants, such a person stops training. This type of equilibrium system might mean satisfying all or just some chosen needs of an athlete, e.g., the need for affiliation. Sport training might be compared with executing a definite profession. Sport, so to say, becomes the tool to satisfy ones needs.

'Professional Amateur' Type of Commitment

Persons belonging to this type of sport commitment dynamics are at the same time highly involved in two kinds of activities, e.g., sport and professional work, sport and family, sport and education or two sport disciplines. An individual subordinates their own functioning to two selected life domains, other activities are supplemental. The two chosen life domains are the major fields of an individual's need fulfillment. People described by this type of commitment participate regularly in practice of high intensity. The trait that distinguishes these athletes from those with a 'professional' type of commitment is that, because of their high commitment to their second activity, they do not allow time for restitution purposes after exertion. While 'professional' athletes take rest practice, those with 'professional amateur'

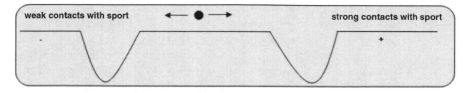

Fig. 5.4 The 'professional amateur' type of sport commitment. The continuum of contacts with sport for an athlete simultaneously involved in sport and other (one) selected activity. The system's equilibrium points are close to the end of continuum describing strong contacts with sport (+) and also to the end describing weak contacts with sport (−). The *ball* symbolizes the dynamic system of an athlete

commitment very often go to work or university or take care of family matters. However, sport will always be a little less important in this arrangement. An individual will never be able to sacrifice work to sport activity.

The system of such an athlete will be characterized by two coexisting attractors, between which a struggle will occur to 'entice' an individual into one or other basin of an attractor. The attractors will be quite deep and narrow (Fig. 5.4).

Research Question

Assuming that it is possible to distinguish the groups of athletes differing in their type of sport commitment, the research problems deriving from the theoretical considerations just presented might be expressed by the following question:

Is the definite type of sport commitment dynamics connected with a specific level of sport achievements?

Method

The aim of the present research was to distinguish the separate types of sport commitment and verify the relation between the type of sport commitment dynamics and sport results achieved by the athlete. The variables are in the form of individual differences. The participants, all athletes, were questioned once, with the use of a personal computer and paper/pencil materials.

Variables and Research Indicators

- The 'sport standard' achieved by participants. The measurement of sport standard was made with a use of a Sport Achievements Survey. The indicator of the variable is a point result obtained by the participant.

- The individual 'sport commitment type' is assigned by competent judges both on the basis of diagnosis of equilibrium states in the Mouse Task and using classic content analysis of the athlete's verbal statements about the involvement of sport in the typical schedule of daily activities. It was possible to assign the athletes into one of the following types of sport commitment dynamics: 'professionalism', 'amateurishness', 'fervor', 'professional amateurishness'.
- Dynamic indicators in the 'mouse task'
    - 'mean distance' – the mean distance of the mouse cursor from the middle of the screen, in pixels.
    - 'speed' – the speed of mouse cursor's movements while executing an assignment, in pixels/s.
    - 'acceleration' – the tempo of change in mouse cursor's movements on the computer screen, in pixels/s^2.
    - 'rest time' – the duration, in which the mouse cursor remained motionless during assignment, in hundredths of a second (s*100).

Participants

The research was conducted on athletes representing diverse sport standards. 67 athletes participated in the study. The group was composed in such a manner that all participants were taking part in sport competitions of different importance. They included members of a national team in different sports (52 %), including the representatives of Poland for the Olympic Games in Athens in 2004. The rest were athletes competing in national contests and members of sports associations, i.e., the Academic Sports Association of the Warsaw School of Social Sciences and Humanities (AZS SWPS) and the Academic Sports Association of the University of Warsaw (AZS UW).

The athletes were training in disciplines that have a measurable method of evaluating the results (points, time, distance), e.g., swimming, judo, fencing, wrestling. In the sample athletes training for individual disciplines were in the majority – 84 %. In the sample 59 % were men, and the mean age of the group was 20 years (SD = 3.1). Most frequently, the athletes declared their workouts to take place 5 times per week (the minimum amount was twice a week, the maximum 13 times).

Research Procedures

Experimental Procedure

The research was conducted by individual appointment. At the beginning, athletes signed informed consent to participation in a 'Me in sports' study. In the first part, participants single-handedly recorded their own answer to the following question:

Tell me, as precisely as you can, how does your typical day look like. Describe, what you most frequently do and how long does it take you, with whom and where do you meet. Focus on the matters most important to you.

After introducing the participants to the instructions for this first part of the experiment, and giving a presentation of how the recording worked, the experimenter left the room while participants recorded their replies. Such action was undertaken to emphasize the anonymity, keep the experiment conditions standard and decrease the influence of other people's presence (in this case the experimenter) during the research. The recording of participants' replies was undertaken with a use of laptop and a microphone. The maximum time for a response was 5 min.

After the participant finished the registration of their responses, the experimenter came back to the room and presented the instructions for the next task. This consisted of the evaluation of a recording of their own reply to the question. This task consisted of two parts: a training part and the proper part. The instructions, presented on the computer screen to the participant before the training part, gave the information that it was a training task to get used to the functioning of the Mouse program. It explained, that the participant would hear a recording of a fake daily activity schedule (not theirs) via the loudspeakers; while listening to it they were to mark to what extent the content of the account was associated with sport. The participant was to do this with a use of computer program. The athlete was informed that two symbols would appear on the screen: a white point in the middle of the screen and a mouse cursor. The participant was to move the cursor on the screen reflecting the current evaluation of activities mentioned in the recording. The stronger the definite activity was connected with sport the closer to the white point the participant was to move the cursor. The weaker the connection the further the cursor was to be moved from the middle of the screen.

The procedure described above concerned the athlete's evaluation of the extent the thing being talked about was connected with sport, according to their own feelings. After resolving all doubts, the participant ran the training part of the program of the Mouse Task, in which they became familiar with the procedure. If any questions occurred, the experimenter answered them. If the participant had no more doubts about the procedure, they were instructed on how to run the program proper. The only difference between the training and the proper version was that in the latter the participant listened to their own recording (their response to the question on their daily schedule of activities mentioned above), not to a fake one. The experimenter left the room for the time the participant undertook the Mouse Task. After finishing, the experimenter came back to the room and gave information about the next phase of the study.

In the last part of the experiment the participant completed the Sport Achievements Survey. Afterwards, the experimenter thanked the participating athlete and answered any extra questions. Going through all procedures took about 30 min.

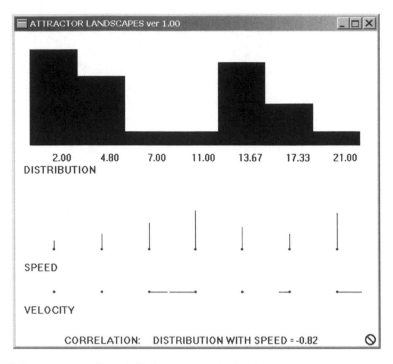

Fig. 5.5 Program output for an individual with two fixed point attractors

Program for Analyzing the Patterns of System Dynamics

To analyze the records of mouse movements on the computer screen, these movements were changed into time rows with a use of the 'Analyser' program written by Nowak and Vallacher (2002). Next, these rows were inserted into a second computer program which looked for the attractors of the dynamic system. Figure 5.5 presents output from the computer program used by Johnson and Nowak to analyze type and number of attractors in a system.

Based on the output of the Johnson-Nowak computer program four competent judges attributed one of five types of attractor to each participant: 1. fixed point attractor on the left side of the output distribution; 2. fixed point attractor on the right side of the output distribution; 3. two fixed point attractors, one on the left side and one on the right side of the output distribution; 4. three or more single point attractors; 5. no attractor (unspecified). Definite combinations of attractors received names which describe types of sport commitment: 'professionalism', 'amateurishness', 'fervor' and 'professional amateurishness'. The final type of each athlete's attractor was a result of all judges evaluations. The criterion was the agreement by three of the four judges. The discrepancies between judges (three cases) were decided by an independent arbiter. The coefficient of concordance Kendall's W for the competent judges was $W(66) = 0.75$; $p < 0.01$.

Content Tools for Assignment of Commitment Types

The categorization of sport commitment types was also conducted using content analysis of participants' statements concerning the description of their typical daily activities. Transcripts of the verbal recordings of the athlete's statements were analyzed in two ways by six competent judges. The competent judges were alumni and students in their last years of Psychology at the University of Warsaw.

Definition Task of Content Analysis

In the first task of content analysis (the Definition Task) judges were asked to assign the dominant type of sport commitment to participants' statements, using the commitment type definitions which were presented to them. The criterion for assigning a particular type of commitment to a participant was agreement between five of the six competent judges. In case of divergences in the judges' rating, the participant was assigned an 'unspecified' type of commitment. In this task the coefficient of concordance Kendall's W for the competent judges indicated $W(66) = 0.64; p < 0.01$.

Survey Task of Content Analysis

In the second task (the Survey Task) the same transcripts were evaluated by another six competent judges. Their task was to complete a survey consisting of 12 statements diagnosing the type of sport commitment with regard to each participant's account. There were three statements designed to be true for each one of the types of commitment and false for the other three. These statements were prepared by the authors and referred to the importance and role the sport played in life of the individual.

In this task, judges were supposed to decide which of the statements were truthful in relation to each participant's accounts. The condition to admit that a definite statement adequately describes a person was agreement between five of the six judges. In all other situations (less than five judges agreed) the statement was considered to be false for the particular person. In this manner, one survey per participant was acquired.

Using the survey it was possible to assign each athlete to a specific type of sport commitment. If two of the three statements designed to hold for one type of commitment were true then this was sufficient to describe a person as characterized by that commitment. When a participant was described by more than one type of commitment e.g., 'amateurishness' and 'fervor', they were marked as having an 'unspecified' type.

Sport Achievements Survey (SAS)

The Sport Achievements Survey (SAS) was designed by the authors to measure objectively the successes in sport of the participating athletes up until that time. The survey consisted of 20 questions, to which the athletes responded 'YES' or 'NO'. The questions relate to objective indicators of sport standard, e.g., winning a medal in the Championship of Poland, breaking the European record, being a member of National Team in the respective sport. A person achieved one point for each positive answer. The survey results could range from 0 to 20 points. Obtaining a high score shows a higher sport standard of the athlete, whereas a low score illustrates lower sport achievements of the participant. The coefficient of inter item reliability, Cronbach's alpha for this method, was 0.876.

Results

The mean duration of the athletes statements was 121 s (SD $= 59$ s), the shortest statement lasted 28, the longest 300 s. To assign sport commitment types to the athletes, three independent measures, all based on the analysis of (transcripts of) the recordings of athletes' statements, were administered using the Mouse Task, the Definition Task and the Survey Task.

The first assignment of sport commitment type was based on an analysis of the computer mouse movement in the Mouse Task, carried out by each participant while listening to their own previously recorded response to the question concerning the role sport played in daily schedule of their activities.. Using the printout of the Johnson-Nowak computer program, four competent judges categorized the participants into one of the four sport commitment types. The numbers of participants categorized into each of the sport commitment types on the basis of the Mouse Task are shown in Table 5.1. Note the low number of participants that had to be assigned to the 'unspecified' type. This indicates that it was usually the case that at least three of the four judges agreed on the commitment type to be assigned.

We also investigated whether the same taxonomy of sport commitment exists when the categorization of participants to a definite type is executed with a use of classic content analysis. In the Definition Task six competent judges, using the transcripts of the recordings of each athlete's response to the daily activity schedule question and given definitions of the four sport commitment types, assigned each participant to one of these types. Table 5.2 presents the number of participants assigned to each of the four types in this Definition Task. The agreement between the six judges was not high; for 22 % of the participants it was not possible for at least five of the six judges to agree, as compared to just 3 % in the Mouse Task. Also the numbers in each type differ from the numbers assigned with a use of the Mouse task.

Table 5.1 The number of participants assigned to each of the four sport commitment types using the Mouse Task

Type of commitment	Number (N = 67)	Percentage (%)
Professionalism	10	14.9
Fervor	16	23.9
Professional amateurishness	28	41.8
Amateurishness	10	14.9
Unspecified	3	4.5

Table 5.2 The number of participants assigned to each of the four sport commitment types using the Definition Task

Type of commitment	Number (N = 67)	Percentage (%)
Professionalism	29	43.2
Fervor	7	10.4
Professional amateurishness	6	9.0
Amateurishness	10	14.9
Unspecified	15	22.4

Table 5.3 The number of participants assigned to each of the four sport commitment types using the Survey Task

Type of commitment	Number (N = 67)	Percentage (%)
Professionalism	25	37.3
Fervor	10	14.9
Professional amateurishness	7	10.4
Amateurishness	10	14.9
Unspecified	15	22.4

In the second of the content analysis tasks, the same six competent judges were asked to complete the questionnaire, using transcripts of the recordings of each participant's response to the daily activities question. Table 5.3 presents the number of participants assigned to each of the four sport commitment types as a result of the Survey Task. As with the Definition task for 22 % of the participants no five judges could agree on the same commitment type. Again the numbers of participants assigned to each commitment type differ from those assigned using the Mouse task.

Comparison of Methods of Type Assignment

After comparing the three methods of assignment to sport commitment type, differences were observed between the number of athletes assigned to each of four types by the Definition Task, the Survey Task and Mouse Task. Consequently, we investigated to what extent the categorizations of sport commitment types were congruent with each other. The results are presented in Table 5.4.

Table 5.4 The agreement on sport commitment types in separate assignment tasks

	Mouse Task	Definition Task
Mouse Task		
Definition Task	30 %	
Survey Task	28 %	65 %

Agreement on the sport commitment type was highest between the two methods of content analysis of the participants statements. Considerably lower agreement was found between the assignments of commitment types using content analysis on the one hand and using printouts from program for patterns analysis of system dynamics (the Mouse Task). The numbers of athletes assigned to the same type of commitment in all three tasks, were, by type: 'professional' 8, 'amateur' 6, 'fervor' 0, 'professional amateur' 0, 'unspecified' 0.

To ascertain the overlap between the three different methods of assignment to sport commitment type, a χ^2 test was administered. An association was to be seen between the commitment types assigned both by Mouse and Definition Tasks, $\chi^2(16, N = 67) = 43.72$; $p < 0.01$, and by the Mouse and Survey Tasks, $\chi^2(16, N = 67) = 43.04$; $p < 0.01$. The agreement between the commitment types assigned by the two content analysis tasks (Definition and Survey) was also statistically significant, $\chi^2(16, N = 67) = 93.04$; $p < 0.01$.

Given the low correlations between the three methods of assigning participants to the four commitment types, we investigated the overlap and differences between them. First, the frequencies for commitment types in Mouse Task and Definition Task are presented in Table 5.5.

Looking closely at the numbers in Table 5.5 one notices that the main confusion is between the assignments to the 'fervor' and 'professional amateurishness' types; on the other hand, both 'professionalism' and 'amateurishness' are given similar assignments by both tasks. Maybe the Definition Task poorly diagnosed the type 'professional amateurishness', placing most people who belonged to this type according to the Mouse Task into the 'professionalism' type. Similarly of the participants assigned to the 'fervor' type by the Mouse Task, by the Definition Task half are assigned to the 'professional' type and the rest to the 'unspecified' type. Maybe the Definition Task is, to a small extent, able to diagnose more complex types, such as 'professional amateurishness'.

The differences in the assignments to commitment types by the Mouse and Survey Tasks were also compared. The best correspondence was between 'professionalism' and 'amateurishness' types (see Table 5.6). Only one of participants assigned to the 'professional amateurishness' type in the Mouse Task, was assigned to this type by the Survey Task; the rest were assigned to all of the other types. Similarly, participants assigned to the 'fervor' type in the Mouse Task were assigned evenly across all the types, except 'amateurishness', by the Survey Task. Therefore the Survey Task of content analysis, just as the Definition Task, diagnoses fewer athletes as being of a type with more complicated dynamics, i.e. 'fervor' and 'professional amateurishness'.

Table 5.5 The number of participants assigned to each of the four sport commitment types in the Mouse and Definition Tasks

Commitment type from Mouse Task	Commitment type from Definition Task					
	Professionalism	Amateurishness	Professional amateurishness	Fervor	Unspecified	Total
Professionalism	9	0	0	0	1	10
Amateurishness	1	6	1	1	1	10
Professional amateurishness	13	4	3	3	5	28
Fervor	6	0	1	1	8	16
Unspecified	0	0	1	2	0	3
Total	29	10	6	7	15	67

Table 5.6 The number of participants assigned to each of the four sport commitment types in the Mouse and Survey Tasks

Commitment type from Mouse Task	Commitment type from Survey Task					
	Professionalism	Amateurishness	Professional amateurishness	Fervor	Unspecified	Total
Professionalism	8	0	0	0	2	10
Amateurishness	1	6	1	2	0	10
Professional amateurishness	11	3	1	4	9	28
Fervor	5	0	4	4	3	16
Unspecified	0	1	1	0	1	3
Total	25	10	7	10	15	67

Finally we present the comparison of the numbers of participants assigned to the four commitment types by both methods of content analysis, the Definition and Survey Tasks – see Table 5.7. The 'professionalism' and 'amateurishness' types were consistently assigned by the two tasks. The 'professional amateurishness' and 'unspecified' types were not assigned coherently; the participants assigned to either one of these types by either one of the two tasks, the other task assigned across all the types.

To evaluate the strength of association between commitment types assigned by the three tasks, a Cramer's V analysis was conducted. Table 5.8 presents the strength of relationship between the assignments made by the three tasks. The strongest relationship is seen between the Definition and Survey Task, both based on content analysis.

Summarizing, the two content based methods gave similar assignments of types to athletes, while the Mouse Task deviated. Mouse Task assigned more athletes to the 'professional-amateurishness' type than to other types, but this type was the least used by the two content analysis methods. The next most frequent type by the Mouse Task was the 'fervor' type, which was the second least used by the content analysis tasks.

Relationship Between Sport Commitment Type and Achievement

The relationship between types of sport commitment and level of sport achievements, being of particular interest, was measured using a one-way analyses of variance. The explicans was athlete's sport standard, as measured by the Sport Achievements Survey (SAS); the explicandum was the commitment type assigned by one of the three tasks: Mouse, Definition and Survey. The actual SAS scores covered the full permissible range of 0 to 20 points, with a mean of 5.6 points (SD = 3.73). Their distribution did not differ significantly from a normal distribution (Kolmogorov-Smirnov test, $D = 1.29$; $p = 0.06$).

We begin by using the assignments from the Mouse Task. Here the 'unspecified' type was excluded since there were only three participants with this type (also it was not part of the theoretical introduction). The analysis of variance confirms the hypothesis concerning the relationship between sport commitment type and SAS scores, $F(3, 60) = 13.25$; $p < 0.01$. The mean SAS scores for separate sport commitment types are presented on Fig. 5.6.

A significant differences in SAS scores for each of the commitment types was found, using Scheffé's contrast, only between the 'professional' type of commitment versus all the other commitment types. For the other three types the differences in SAS scores were not statistically significant.

Analogous statistical analyses were conducted on the two content analysis tasks. However, due to the higher number of participants assigned to the 'unspecified' commitment type, this type was here included in the analysis.

Table 5.7 Table of frequencies observed for categorization in Definition Task and Survey Task

Commitment type from Survey Task	Commitment type from Definition Task					
	Professionalism	Amateurishness	Professional amateurishness	Fervor	Unspecified	Total
Professionalism	23	0	1	0	1	25
Amateurishness	0	9	1	0	0	10
Professional amateurishness	2	0	2	1	2	7
Fervor	0	0	0	3	7	10
Unspecified type	4	1	2	3	5	15
Total	29	10	6	7	15	67

Table 5.8 The strength of relationship between commitment types assigned by the three tasks: Mouse, Definition and Survey Tasks

	Mouse Task	Definition Task
Definition Task	V = 0.42; p < 0.01	
Survey Task	V = 0.39; p < 0.01	V = 0.60; p < 0.01

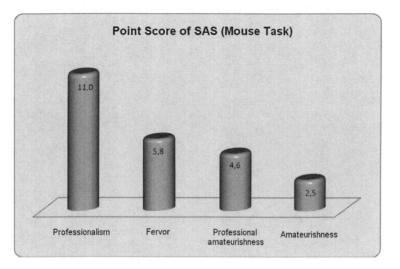

Fig. 5.6 The comparison of Sport Achievements Survey scores for athletes for each sport commitment type as assigned by the Mouse Task

Using the commitment type assigned in the Definition Task as the explicandum, an analysis of variance again confirms the hypothesis concerning the relationship between sport commitment type and achievement as measured by the SAS $(F(4, 62) = 11.74; p < 0.01)$. The mean SAS results for separate sport commitment types from the Definition Task are presented in Fig. 5.7.

The variances within each type do not differ significantly (Levene's $F(4, 62) = 1.48; p = 0.21$). The mean SAS score for the 'professional' commitment type differed significantly, using Scheffé's contrast test, from other commitment types, except for the 'professional amateurishness' type.

Using the sport commitment types as assigned in the Survey Task as the explicandum, the analysis of variance also confirmed the relationship between sport commitment types and achievement from the SAS survey, $F(4, 62) = 14.45; p < 0.01$. The mean SAS scores for separate sport commitment types assigned in the Survey Task are presented on Fig. 5.8.

The Games-Howell contrast test showed that there were statistically significant differences in SAS scores between the 'professional' commitment type and the 'fervor' and 'amateurishness' types, and between the 'amateurishness' type and the 'professional amateurishness' and 'unspecified' types, as well as between the 'fervor' and 'unspecified' types.

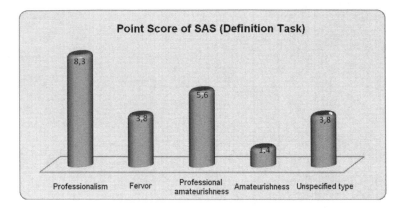

Fig. 5.7 The comparison of Sport Achievements Survey scores for athletes divided in groups according to dynamic types of sport commitment (Definition Task)

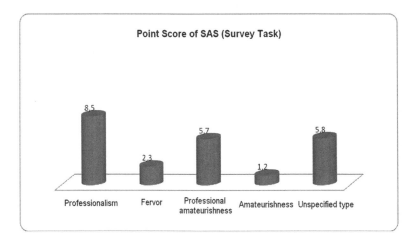

Fig. 5.8 The comparison of Sport Achievements Survey scores for athletes divided in groups according to dynamic types of sport commitment (Survey Task)

Mouse Movement and Sports Achievement

We turn next to the relationship between Sport Achievements Survey scores and dynamic variables from the Mouse Task, as shown in Table 5.9. The strongest correlation with the SAS survey result occurred in case of the 'mean distance' variable: people achieving higher sport results moved the mouse significantly closer to the centre point of the screen while listening to their own previously recorded reply. The 'rest time' variable had a weak positive correlation with sport standard: the higher score achieved in SAS the longer was the 'rest time' of the mouse cursor. The remaining dynamic variables showed no statistically significant correlation with SAS scores.

Table 5.9 The correlation matrix for dynamic variables in Mouse Task with 'sport standard' variable

	Mean distance	Speed	Acceleration	Rest time
SAS score	$r = -0.43$	$r = -0.05$	$r = -0.04$	$r = 0.21$
	$p < 0.01$	$p = 0.34$	$p = 0.36$	$p < 0.05$

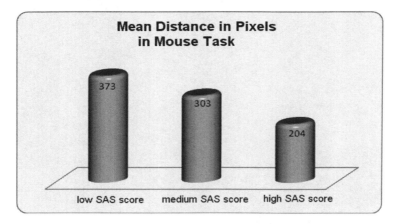

Fig. 5.9 The comparison of mean distance in Mouse Task for athletes divided according to low, medium or high SAS score

This relationship between dynamic variables from Mouse Task and sport achievement was checked using one-way analyses of variance. The dependent variables were the Mouse Task variables: mean distance, speed, acceleration and rest time; the independent variable was the SAS score divided into three groups.

The Mouse Task variable mean distance decreased as SAS scores increased ($F(2, 64) = 6.43$; $p < 0.01$, see Fig. 5.9). There was a significant difference of mean distance, using the Scheffé's contrast test, between low and high SAS scores.

The rest time increased with SAS scores ($F(2, 64) = 3.74$; $p < 0.05$, see Fig. 5.10). The difference in rest times between high and medium SAS scores was significant, again using Scheffé's contrast test.

The strength of relationship between 'sport standard' and the Mouse Task dynamic variables was also checked with the screen divided into two spheres: close to and far from the centre of the screen. The coefficients with In the sphere close to the centre of the screen sport standard was significantly correlated with some dynamic variables: 'speed' ($r = -0.26$; $p < 0.01$), 'acceleration' ($r = -0.24$; $p = 0.047$) and 'rest time' ($r = 0.35$; $p < 0.01$). However, in the ring further from the centre of the screen sport standard was correlated only with the 'mean distance' ($r = -0.32$; $p < 0.05$). As might be expected, the participants with higher sport standard moved the mouse cursor quickly in the outer ring but once in the inner ring moved more slowly, with smaller acceleration and kept the cursor motionless significantly longer.

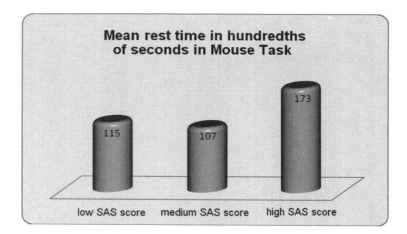

Fig. 5.10 The comparison of mean rest time in Mouse Task for athletes divided according to low, medium or high SAS score

Discussion

The aim of this research was to ascertain whether or not an athlete's type of sport commitment is a factor determining their sport success. From the theories we reviewed, commitment is likely to be superior to a system of factors: athlete's characteristics, coach characteristics and the influence of external components, each of which affects the others, and in turn affects the achieved successes in sport. Applying classic methods of content analysis, as well as tools arising from complex systems theory, it was usually possible to assign one of the four types of sport commitment described in theoretical introduction to the athletes participating in the study (an 'unspecified' type was occasionally needed).

The assignment of commitment type using the mouse paradigm was based on an analysis of auto-reflection by the athletes listening to a recording of the schedule of their own typical daily activities. We assumed that, similarly to an auto-reflection stream in reference to global characteristics of 'I' dynamics (see Krejtz 2003; Vallacher et al. 2002), in the spoken statements the dynamics of the strength of relationship between daily activities and sport would lead to definite stable patterns or attractors. Using printouts from the program for analyzing these dynamics it was possible for expert judges to assign each person to one of the four sport commitment types. A high degree of agreement between the judges was found for these assignments.. It is the first time that this tool has been used to examine the dynamic patterns of sport commitment.

It was also possible to assign sport commitment types to the athletes using classic methods of content analysis of transcripts of the participants statements. However compared to the Mouse Task, there was less agreement between judges, using either the Definition Task or the Survey Task, in unequivocally assigning a commitment type.

All three ways of assigning commitment type are based on an analysis of the athlete's response to the question concerning their daily schedule of activities. However, the manner of analyzing the material was different. The Definition and Survey Tasks used a classical assignment method of content analysis with expert judges analyzing transcripts of recorded statements with criteria fixed beforehand. In the Mouse Task it is a precise mathematical analysis of an evaluation of the response of the athlete reacting to this same recorded statement: the assignment of commitment type is made with the use of printouts illustrating the number of attractors in the participant's system. The number of possible solutions of attractor systems is not defined in advance, so this tool is able to describe a more extensive range of types. Whereas in classic studies of content analysis, since categories are defined in advance, judges may omit giving a more description that is more appropriate, as this is not an option that is offered them. A further advantage in using the mouse paradigm is the increased objectivity; the material itself does not interfere with opinions, experiences and knowledge of competent judges, since they evaluate the printouts from the program for the analysis of system dynamics.

Additionally, only the mouse paradigm provides dynamic accounts which may then be analyzed. It is especially visible in case of statements with high dynamics as in 'fervor' or 'professional amateurishness' commitment types. Thus, diagnosis of commitment types using the Mouse Task captures information which is absent from the classic content analysis. Maybe this is why in the content analysis tasks there were fewer athletes assigned to the commitment types: 'fervor' and 'professional amateurishness', whereas there were more athletes assigned to the 'unspecified' or the 'professional' types.

On the other hand, the mouse paradigm might have the disadvantage that its results do not contain specific psychological content (except for information about the configuration of system attractors), whereas classical content analysis does. So these methods complement one another; the mouse paradigm is a more precise method and the classic content analysis can capture more of the psychological content (see Ziembowicz 2007).

The null hypothesis of no relationship between the different methods for assigning athletes to commitment types was rejected. However, the strengths of relationships between separate tools was not very high, especially not between the Mouse Task and the content analysis tasks. Both methods of content analysis are based on the same material, that is the transcript of the athletes accounts of the activities, whereas the Mouse Task is based on the dynamics of mouse movements made by the athletes while listening to their own accounts. So maybe the mouse paradigm detects information that is not captured by classic content analysis.

The relationships between the dynamic variables captured by the Mouse Task and the level of sport achievements measured by Sport Achievements Survey is interesting from a complex systems theory perspective. While executing the Mouse Task athletes with a high SAS score kept the cursor significantly more often close to the middle of the screen, indicating a strong relationship between the activity being described and sport. When these athletes did move the mouse cursor far from the centre half of the screen, they moved the mouse significantly less far away from the centre than athletes with lower sport results, which means that they rarely spoke about activities that were very weakly connected with sport.

Not only the position of the cursor but also its movement in Mouse Task varied between athletes with different SAS scores. The athletes with high SAS scores more rarely changed the position of the cursor on the computer screen than those with middle SAS scores; they focused on one topic longer. In contrast, there was no significant difference in mouse rest times between athletes with high and low SAS scores; maybe, amateurs as well as professionals more rarely changed the type of activities described by them. There was also a negative relationship between the cursor's acceleration in the sphere close to the centre of the screen and the athletes' SAS score; athletes with a high SAS score modified their focus of attention less while discussing their daily activities.

In a complex system low speed and acceleration values co-occur with a higher value of the system's rest time close to an attractor. Athletes with high SAS scores met this criterion most of the time; demonstrating the existence of an attractor in the area of strong sport commitment. These results indicating the presence of an attractor in an area strongly connected with sport are based on objective mathematical measures rather than on the subjective evaluation of competent judges about the existence of such a type of attractor. An additional advantage of these results is that they illustrate in detail the time path of the stream of auto-reflection about the activities undertaken by an individual.

Another variable which was significantly connected with the sport standard of the athletes was the mean distance of the mouse cursor from the middle of the screen in the Mouse Task. The score obtained by the athletes in SAS negatively correlates with the mean distance of cursor from the centre point of the screen. Athletes who achieve high sport results, move the mouse cursor a smaller distance during the task, showing that most of the time in their typical schedule of daily activities is spent on actions connected with sport. It is reasonable that such athletes really do spend more time during the day on sport, and that they conjoin more potentially neutral activities with their sport activity, e.g., the friends they talk to also do sports training. The essence of the sport commitment model presented here is precisely the inclusion by a highly committed individual of the off-sport activities with sport training and organizing their own life around it, creating a sport oriented dynamic system.

Particularly significant from the complex systems theory point of view, are the results of the correlation analysis between sport standard and the dynamic variables from the Mouse Task with the screen divided into two areas: close to and far from the middle of the screen. The coexistence of smaller speed and acceleration of the system with longer rest time of the mouse cursor in the area close to the middle of the screen for athletes achieving higher SAS scores, indicates the existence of a strong attractor connected with sport. In the area far away from the middle of the screen speed, acceleration and rest times are not significantly connected with sport standard: there is a lack of attractor 'inhibiting' these best athletes from sport.

Final Conclusions

This research is a new way of approaching the understanding of athletes' sport commitment. Previous theories concerning the degree of commitment of athletes to sport and their performance lacked the specification of commitment dynamics. Hitherto, the commitment has been treated as a variable, which might be described on a continuum, where one end signifies low, and the other end high commitment. Grouping sport commitment into four independent types has allowed us to comprehend the difference in results obtained by the athletes in different commitment types. For the first time, a direct relationship has been shown between the type of sport commitment and the sport results achieved by the athletes.

So we can justifiably claim that sport commitment plays a key role in achieving high sport results. It is not simply the level of commitment itself, but the pattern of commitment dynamics of the athlete, that is significant. The commitment type determines the amount of time, financial resources and energy, which an individual athlete allocates to sport activities, defining at the same time the sort and extent of needs fulfilled through sport. It is also the over-arching factor of the many elements, all connected with each other, that might have an influence on sport results achieved. The type of commitment has emergent properties. This especially holds for the 'professional' type of commitment; a high standard in sports requires a very subtle combination of factors. Quite often, the differences between the configuration of psychological elements of athletes described as having a 'professional' type of commitment and 'professional amateur' or between 'professional' and 'fervor' is very little, but this difference determines whether or not the athlete stays in sport and achieves successes. In other words, it might be difficult to distinguish athletes during practice, but the differences between them will be visible when looking from a broader perspective and observing the role of sport in their whole life.

The precise definition of different types of sport commitment, their relationship with sport results and the mechanisms of keeping the athletes in highly-qualified sport requires further investigation. It seems, though, that looking at athletes from the perspective of their sport commitment type might allow specific recommendations to be given to a coach; for instance the most appropriate method of working with an athlete might depend on the pattern of their sport commitment dynamics. The theory we have presented could be useful in the analysis of other similar activities, e.g., creative work, musical activity or scientific work.

References

Anshel, M.H.: Sport Psychology. From Theory to Practice. Benjamin Cummings, San Francisco (2003)

Gernigon, C., d'Arripe-Longueville, F., Delignieres, D., Ninot, G.: A dynamical system perspective on goal involvement states in sport. J. Sport Exerc. Psychol. 26(4), 572–596 (2004)

Johnson, S.L., Nowak, A.: Dynamical patterns in bipolar depression. Pers. Soc. Psychol. Rev. **6**, 380–387 (2002)

Krejtz, K.: Dynamika Ja. Regulacyjne funkcje globalnych własności Ja w strumieniu autorefleksji. PhD thesis, Psychology Department, University of Warsaw (2003)

Lesyk, J.: Developing Sport Psychology Within Your Clinical Practice. Jossey-Bass, San Francisco (1998)

Morrow, P.: The Theory and Measurement of Work Commitment. JAI Press, Greenwich (1993)

Nowak, A., Vallacher, R.: Dynamical Social Psychology. Guilford Press, New York (1998)

Nowak, A., Vallacher, R.R.: The development of software for identifying attractors in time series. Unpublished research data (2002)

Vallacher, R., Nowak, A.: Dynamical social psychology: finding order in the flow of human experience. In: Kruglanski, A., Higgins, E. (eds.) Social Psychology: Handbook of Basic Principles. Guilford Publications, New York (2007)

Vallacher, R., Nowak, A., Froehlich, M., Rockloff, M.: The dynamics of self-evaluation. Pers. Soc. Psychol. Rev. **6**, 370–379 (2002)

Williams, J.M.: Personality characteristic of successful female athletes. In: Straub, W.F. (ed.) Sport Psychology: An Analysis of Athlete Behavior, 2nd edn, pp. 353–359. Mouvement, Ithaca (1980)

Ziembowicz, M.: Paradygmat myszy. In: Winkowska-Nowak, K., Nowak, A., Rychwalska, A. (eds.) Modelowanie matematyczne i symulacje komputerowe w naukach społecznych, pages, pp. 27–31. Wydawnictwo SWPS Academica, Warszawa (2007)

Chapter 6
Conflict and the Complexity of Cognitive Processes

Katarzyna Samson

Abstract From the perspective of Dynamical Systems Theory, the collapse of complexity is one of the crucial phenomena involved in conflict escalation and intractability. It happens when the relations between distinct psychological or social elements involved in conflict become aligned and begin to mutually reinforce each other. At an individual level, a collapse of complexity is reflected in a decrease in the complexity of cognitive processes – the dimensions on which reality is perceived converge, so, at the end of the day it all comes down to being **either with us or against us**. The first part of this chapter outlines a number of different constructs related to the complexity of cognitive processing, describes their most common measures, and briefly presents their ontogenesis. In the second part, the empirical evidence for the impairment of cognitive capacities in a situation of social conflict is shown. Finally, the Dynamical Systems Theory approach to conflict is briefly presented; focus is placed on the role of complexity, in the processes both of conflict escalation and of conflict resolution.

A couple of years ago, I spent the winter rock climbing in Patagonia. I had a great climbing partner and friend; we knew each other forwards and backwards. He was witty, outgoing and extremely charming. His vast experience as a mountain guide promoted trust, while his extraordinary social skills made spending time with him pure fun. Although he was sometimes presumptuous and even insolent, I truly admired his climbing skills, motivation and persistence. I also loved that he respected my independence and strong opinions, and felt a bit flattered to be called guapa, which in Spanish means beautiful. One evening, after a long day of climbing, we were heading back towards the campground. We were finishing the route in really bad conditions, a couple hundred meters before our summit a storm had hit us; we both were cold, hungry and exhausted. We stopped just to figure out

K. Samson
Faculty of Psychology, University of Warsaw, ul. Stawki 5/7, 00183 Warsaw, Poland
e-mail: ksamson@psych.uw.edu.pl

A. Nowak et al. (eds.), *Complex Human Dynamics*, Understanding Complex Systems,   93
DOI 10.1007/978-3-642-31436-0_6, © Springer-Verlag Berlin Heidelberg 2013

our whereabouts, and a small difference of opinion concerning directions suddenly turned into a full-blown fight. For a while we were both pointing out each other's faults and throwing insults around. Then he announced that, after all, I wasn't even pretty, and walked away.

This story is an example of how conflict can develop in any area of our functioning, including basic cognitive processes. What was it, exactly, that caused the turnaround of my partner's perceptions? Can Psychology explain this behavior? Why does a minor conflict result in such major consequences? In the current chapter a number of different constructs related to the complexity of cognitive processing are introduced, along with their most common measures. Developmental theories are also described. After this, a brief overview of studies linking complexity of cognition to conflict is presented. Finally, the approach to destructive conflicts taken by Dynamical Systems Theory is discussed, along with insights this brings to conflict resolution.

The Complexity of Cognitive Processes

The complexity of a cognitive process refers to a number of related constructs associated with the ways in which people think, decide and relate to each other. Different authors have given somewhat different interpretations to this concept, but in general they reflect the belief that some people employ fewer dimensions when perceiving and describing reality, and tend to make only the most blatant discriminations between different dimensions, while others tend to use more dimensions and make fine discriminations between them. The complexity of cognitive processes is essentially context free and refers only to the structural characteristics of information processing, which are placed on a simple to complex continuum.

> At the simple end of the continuum, decisions are characterized by anchoring around a few salient reference points; the perception of only one side of an argument or problem; the ignoring of subtle differences or similarities among other points of view; the perceiving of other participants, courses of action, and possible outcomes as being either totally good or totally bad; and a search for rapid and absolute solutions in order to achieve minimization of uncertainty and ambiguity. At the complex end, we find flexible and open information processing; the use of many dimensions in an integrated, combinatorial fashion; continued search for novelty and for further information; and the ability to consider multiple points of view simultaneously, to integrate them, and then to respond flexibly to them (Suedfeld and Tetlock 1977, p. 172).

The roots of this family of concepts can be traced back to Adorno's theory of authoritarianism (Adorno et al. 1950), in which he describes the tendency for perceiving social reality in an over-simplified way which results in prejudice, intolerance, excessive conformity, superstitions and stereotyped thought patterns. The best known constructs that refer to this structural characteristic of information processing are cognitive structure (Scott et al. 1979), conceptual complexity (Schroder et al. 1967), and integrative complexity (Suedfeld and Tetlock 1977; Suedfeld et al. 1992).

The term 'cognitive complexity' was introduced by James Bieri (1955), who assumed that each individual possesses a system of personal constructs (Kelly 1955) through which they perceive their social world. Constructs represent the environment using bipolar perceptions (such as happy–sad, or intelligent–stupid), which serve to characterize objects cognitively. For Bieri, the degree of differentiation of the construct system is a formal characteristic of a cognitive structure – it reflects its cognitive complexity or simplicity. A system of constructs which permits an accurate differentiation between people, is complex. A system which differentiates between them poorly is considered cognitively simple in structure. Cognitive complexity is thus a characteristic, reflecting an individual's ability to record perceptions of their social world in a multidimensional manner and which then serves as a basis for formulating beliefs and opinions.

More recently, individual differences in cognitive competences have been described in terms of conceptual (Schroder et al. 1967) or integrative (Suedfeld and Tetlock 1977; Suedfeld et al. 1992) complexity. In these approaches, the complexity of cognitive processes is a function of a person's differentiation and integration abilities. Differentiation refers to the number of different dimensions or attributes of stimuli, which are recognized and taken into account in decision making. More complex information processing requires the recognition of a larger number of characteristics present in a multidimensional stimulus. Differentiation is a necessary, but not a sufficient, prerequisite for integration, which refers to the character (independence, hierarchical interaction, multiple patterns) of conceptual connections between differentiated dimensions. At the simple end of the continuum, there is no tolerance for ambiguity, only one side of the problem is seen, and the perceptions of other participants or possible outcomes of a situation are seen as black or white. At the complex end, in contrast, multiple points of view are integrated and all subtleties of the situation are considered, while information processing remains flexible and open.

The two concepts, integrative and conceptual complexity, are essentially very similar; the main difference between them is the relative emphasis given to internal processes versus situational factors as determinants of the level of complexity. Conceptual complexity is treated as a personality trait of cognitive style, i.e., people differ along the dimension of simplicity–complexity in their decision making, while integrative complexity of information processing varies not only between individuals but also between situations for each individual. Later versions of integrative complexity show increasing interest in environmental mediators between potential (i.e., trait) and exhibited (i.e., state) complexity. It has been shown that situational variables, such as motivation, threat to basic values, exposure to prolonged stress or time pressure, fatigue, information overload, uncertainty and ambiguity, can temporarily modify the integrative complexity of cognitive functioning, for example, reducing information search, increasing in-group conformity, or promoting stereotyped responses and attitudes (Suedfeld 1980; Suedfeld et al. 1992). Moreover, integrative complexity can also be changed permanently, as a consequence of certain experiences, such as parental influences, or by organizational climates that foster a particular level of complexity, such as the army (Suedfeld et al. 1992). Thus the complexity of cognitive processes may be a personal trait, though not necessarily an unchangeable one. However, the bulk of research focuses on state complexity, which may change in response to factors such as fatigue, stress, or motivation. Whether such changes in the level of cognitive complexity constitute unconscious adjustments to circumstances, or deliberate adaptations remains unclear.

It must be emphasized that the complexity of cognitive processes, whether referred to as cognitive, conceptual or integrative complexity, is a descriptive, not an evaluative dimension (Guttieri et al. 1995). In general, no particular level of complexity is better per se; what is preferable always depends on the characteristics of the situation. Under many circumstances, the ability to process information in a complex manner can be advantageous, especially in situations characterized by non-zero-sum payoff structures. However, under some circumstances, e.g. when in an immediate life threatening situation or faced with an implacable foe, high cognitive complexity proves neither necessary nor beneficial. Complex strategies are usually more costly, in terms of time and resources, than simple ones, while their increased efficacy is doubtful when it comes to zero-sum situations, for example. Thus it seems advisable to adjust the complexity of one's information processing to situational requirements.

Development of Cognitive Capacities

Another different theoretical perspective on the complexity of cognitive processes is the developmental framework represented by such authors as Jean Piaget (1964), Lawrence Kohlberg (1958, as cited in Crain 1985) or Shawn Rosenberg (1988). In this approach it is assumed that cognitive capacities are not given once for all, but develop according to certain universal dynamics. Different cognitive capacities appear in the course of individual development in a fixed order allowing an individual to acquire a more adequate and complex understanding of the surrounding reality; achieving simpler capacities is a necessary prerequisite for the development of more complex ones. In this perspective, people differ in their cognitive capacities because they represent different levels of cognitive development. Attaining the most complex cognitive capacities is thus a developmental potential of every individual, although not everyone realizes it.

Piaget (1964) believed that development is about perfecting adaptation to reality. He viewed humans as active creators of their intellectual development; an individual is constantly interacting with the environment and actively seeking correspondence between their existing organization of knowledge and new environmental experiences. In the adaptational processes of assimilation (including new objects into pre-existing mental structures) and accommodation (changing mental structures to provide consistency with external reality), individuals strive to attain higher levels of development, which gives them a better understanding of the surrounding world. Adjusting to environmental demands takes individuals through four universal stages of development – sensorimotor, pre-operational, concrete operational and formal operational – each at their own pace.

In the cognitive domain, Piagetian development can be understood in terms of acquiring and mastering decentration skills (see Golec 2002b). At different stages of development, individuals overcome different forms of egocentrism, each of which is indispensable for properly adapting to the surrounding reality. At the sensorimotor stage, a child overcomes radical cognitive egocentrism and develops the ability to differentiate between self and other objects. Later, at the pre-operational stage, thinking becomes independent of current activity and a child starts developing an internal representation of the world that allows them to describe objects, events, and feelings. Still, the world at this stage is viewed entirely from the child's own perspective. The concrete operational stage is when interpersonal decentration takes place – a child acquires the ability to differentiate between self and others, not only in terms of physical independence, but also in terms of differentiating between their own and others' perspectives and emotions, though not yet understanding them. Not until the formal operational stage does an individual develop abstract thinking, deductive reasoning and systematic planning, which allows for simultaneously considering different perspectives as well as comparing points of view. One of the main achievements of this stage is the ability to comprehend the relativity and subjectivity of their own reasoning and the impossibility of finding simple criteria of objectivity (Chandler 1978, as cited in Golec 2002b). Throughout the whole development process, individuals attempt to construct more and more accurate and functional models of their surrounding world. However, the highest achieved level of complexity of cognitive functioning differs from person to person; many adults will never learn to think formally.

Inspired by Piaget's cognitive development theory, Shawn Rosenberg (1988) formulated a theory of the development of political reasoning. After Piaget, he assumed that cognitive development serves adaptive functions, i.e. adequate reasoning and functioning in a socio-political reality. Also, that the development of reasoning is governed by universal dynamics, which leads individuals, each at their own pace, through universal cognitive stages. However, his perspective diverged from Piaget's in its view of social environment, to which he attributed a central role in the individual's cognitive development. He claimed that the structure of social environments differs dramatically from person to person (e.g. compare the environment of the president of a company that owns a factory, with that of a line worker in that factory), and that inter-individual differences in attained stages of development are a product of real differences in the environments to which those individuals are exposed. Rosenberg's stages of development (sequential, linear and systematic) are defined by formal characteristics of the organization of knowledge, not its content, so advancing from one stage to another requires a qualitative reorganization of existing structures. Cognitive development is essentially (1) progressing from concrete to abstract reasoning, (2) decentration, and learning to coordinate multiple perspectives in interpreting events and (3) how norms and rules, as well as relationships between them, are understood.

In the **sequential** phase, abstract thinking is absent, and perception and understanding is concrete and proximate. Interpretation of reality is subjective and egocentric, and taking the existence of the perspectives of others into account is impossible. People at this stage are able to track objects, or perceive and describe sequences of events, but without grasping general patterns or relations between them. Behavior is motivated by anticipation of gratification or punishment avoidance; norms or general standards of behavior do not exist.

Linear thinkers are, to some extent, able to build abstract categories, yet possess limited ability to make generalizations. At this stage, it is sequences of activity that are analyzed, and understanding is based on experience, not reflection. The existence of different perspectives appears; however decentration is not yet developed and therefore taking a perspective of another person, or comparing it with one's own point of view, is still impossible. Evaluation is unequivocal and black and white, while behavioral choices are based on external standards, believed to be absolute, and follow authorities.

Only **systematic** thinkers have a high ability of abstract thinking and are able to reflect on both subjective as well as objective conditions of phenomena and their relationships. Not only can they differentiate between perspectives, they are also able to juxtapose relationships between actions and beliefs, as well as coordinate different points of view. Achieving the systematic reasoning stage requires the ability to take the perspective of an external observer and learn about all aspects of a situation. With respect to behavioral norms, relativity is accepted and surpassed, i.e., a moral choice is one which complies with generally accepted values that promote the harmony of social systems. Evaluation thus occurs from an integrated perspective, which allows for the most complex and multidimensional understanding of a situation, and the choice of most adequate attitudes. "The systematic thinker's conception ... extends beyond the limits of his own understanding" (Rosenberg 1988, p. 157).

Advancing from sequential, through linear, to systematic reasoning phases means learning to liberate oneself from an egocentric perspective and to coordinate different points of view. The highest level of reasoning that is attained defines one's cognitive functioning – the more advanced the reasoning stage, the more complex the ways in which sociopolitical reality is perceived and described.

Measuring Cognitive Complexity

There are as many ways to measure the complexity of information processing as there are conceptualizations. To calculate an index of complexity of information processing, archival samples of speech, written material, free-response descriptions or constructs generated to organize stimuli are used. The general idea behind most indices is to find the number of dimensions a person uses in describing their world, and to determine the relationships between them. However, research on the relationship between cognitive complexity and other variables has shown that using different measures of complexity doesn't always yield consistent results (Seaman and Koenig 1974). Taken to the extreme, Erwin and colleagues (1967, as cited in Seaman and Koenig 1974) report a negative correlation between two cognitive complexity measures. Early findings indicated that no single dimension could be posited as underlying the correlations between different measures of cognitive complexity (Vannoy 1965), and it has been hypothesized that a number of distinct tendencies add up to what we call 'cognitive complexity', and thus no single measure captures them all.

The majority of cognitive complexity measures fall into one of two categories: (a) those calculating the number of constructs the subject uses to differentiate between stimuli (Role Constructs Repertory Test and its derivatives); and (b) measures which can be applied to samples of writing or speech, either archival or generated for the purpose (textual analysis measures). Only isolated examples of other measures can be found in the literature. Choosing which to use depends primarily on the researcher's theoretical approach and on the type of data available. However, it is generally advisable to secure measurements using several different procedures at the same time. Exemplary measures from each category are briefly described below.

Role Constructs Repertory Test and Its Derivatives

Kelly's (1955) original version of the Role Construct Repertory Test was developed as an instrument for the elicitation of personal constructs, but it has become the classic tool for measuring the complexity of cognitive processes. In the RCRT procedure the participant first names a number of people, called **elements**, likely to be of some importance to their life. These people are named in response to certain suggestive categories, such as 'An admired friend'. The participant is then asked to consider three of these at a time, and decide in what important way two of them are alike and differ from the third. This way, a number of **personal constructs** – basic terms used to make sense of the elements – are formed. For example, if 'Mum' and 'My spouse' are relaxed, but 'Dad' is rather tense, the construct that the subject uses when thinking about these elements is **tense versus relaxed**. This procedure is repeated for different triples of elements. After a number of constructs has been generated, the participant is asked to go through each construct and mark other elements to which they also apply. This procedure yields a matrix representing how the participant perceives and differentiates his social environment. By comparing combinations of elements applicable to each of the constructs, their similarity is calculated, i.e., if two constructs differentiate between the same elements, they are treated as functionally equivalent. In its original form measuring cognitive complexity using Kelly's grid comes down to calculating the number of generated personal constructs that are independent of each other. The larger the number of independent constructs, the higher is the participant's cognitive complexity.

A number of other complexity measures can also be derived from the grid obtained by the RCRT procedure. For example, Bieri et al. (1966, as cited Seaman and Koenig 1974) proposed a **total complexity score**, which is calculated by the inverse of construct complexity proposed by Kelly. It reflects the number of tied ratings for each element across all constructs; a person that uses the constructs to construe different role figures in an identical way would generate a lot of tied ratings and score low on cognitive complexity. Similarly Fiedler (1967, after Seaman and Koenig 1974) derived **most preferred person (MPP), least preferred person (LPP)** and **assumed similarity of opposites** (the difference between MPP and LPP) scores, which could also be interpreted as measures of cognitive complexity in the sense that attributing positive ratings to people one does not like, or low esteem to people they like, requires attributing both positive and negative qualities to one object, which reflects highly complex cognitive processes (Mitchell 1970 as cited in Seaman and Koenig 1974).

An idea similar to RCRT lies behind Scott's Object Sorting Test (1962) and Crockett's Role Category Questionnaire (Crockett 1965; O'Keefe et al. 1982). In the Object Sorting Test the participants are given a list of words, and asked to repeatedly group them in as many ways as they wish, and then to label the groups. When two groups contain identical words, they are deemed to represent the same attribute, while no overlapping membership represents antithetical dimensions. Maximum independence of two dimensions (labels) is when each pair of groups from different dimensions have exactly half of their members in common. The number of created groups and the relationships between them form the basis for deriving measures of cognitive complexity. In the Role Category Questionnaire the participants are asked to provide free-response descriptions of several persons known to them, e.g. in the most common 'two-peer' version of the RCQ the participants are asked to write two brief essays – one describing a person they know well and like, and another about a person they know well but dislike. These descriptions are then coded for the number of interpersonal constructs they reflect, and the resulting number of constructs is viewed as an index of interpersonal cognitive complexity.

Textual Analysis Measures

Textual analyses are usually conducted on archival written material from the media or public speeches. The number of paragraphs needed from each particular condition (such as source, author or time period) and the selection method used are determined in advance. All information that could be used to identify the condition is removed, and the material is then analyzed by independent scorers. Alternatively, participants can write an essay on some complicated topic or fill in the Sentence/Paragraph Completion Test (Schroder et al. 1967; Suedfeld 1980), in which they are asked to write a number of brief essays based on a stem referring to some socially important issue (e.g. 'abortion' or 'relations to authority') or addressing important domains of the decision-making environment ("When I don't know what to do . . ."; "Rules . . .").

Different researchers use different methods for scoring textual material. For example, Hermann's (1980) conceptual complexity measure employs a simple frequency based automated linguistic analysis of content. It utilizes two dictionaries – one of words indicative of complex cognitive processing, such as 'possibly', 'perhaps' or 'sometimes', and the other of words providing evidence for simple cognitive processing, e.g. 'always', 'never', or 'absolutely'. The ratio of complex processing words to simple processing words is the utterances final complexity score. Other theoretical approaches don't necessarily apply such straightforward content-coding rules. When scoring for integrative complexity (Baker-Brown et al. 1992), skilled coders look for evidence of differentiation and integration in the texts, and rate sections focused on one idea on a scale ranging from 1, indicating no evidence of either differentiation or integration, and relying on uni-dimensional, value-laden and evaluatively consistent rules of information processing, to 7, which indicates high differentiation and high integration. With this method, the system focuses on structure rather than content of the utterances, so almost any material can be scored, from personal letters, through interviews, to political speeches.

Other Measures

In the Political Prediction Test (Sidanius 1978) participants are asked to estimate the degree of political rioting and murder likely to occur in unidentified countries on a 1 (very low) to 10 (very high) scale, based upon six pieces of information: equality of income, national wealth, public health expenditure, military expenditure, voting participation, and proportion of minority groups. The values of the stimuli selected correspond to values found in the real world. In this approach, cognitive complexity is measured as the number of stimulus variables correlating with the levels of the predicted target variables (political rioting and murder), so the greater the number of stimuli taken into account in this task, the greater the participant's cognitive complexity. The political prediction test focuses primarily on the integration aspect of cognitive complexity, and therefore its results are only weakly related to other complexity measures (van Hiel and Mervielde 2003). However, it seems to be the most innovative test of cognitive complexity after Kelly's RGRT and unlike other methods, its administration is simple and calculating the results is fairly straightforward.

Cognitive Capacities in a Situation of Conflict

Being involved in a conflict doesn't leave one's cognitive functioning unaffected. This phenomenon is of great importance for an adequate understanding of the course of conflicts, since cognitive capacities shape a decision maker's understanding of the situation of conflict, and thus affects their attitudes and actions. Psychological theories claim that in a situation of conflict universal mechanisms shaping individual behavior result in simplified, based on dichotomous divisions, perceptions of social reality (Deutsch 1973; Guttieri et al. 1995).

The main factor held responsible for this impairment of cognitive capacities in situations of conflict is emotion. Conflict evokes intense emotions as soon as the distinction between in-group and out-group is drawn, and a threat from the out-group is perceived (Deutsch 1973). The disruptive influence of emotions on cognitive processing can be captured in the simplest terms by the Yerkes-Dodson laws (Zimbardo 1992), which state that the relationship between the levels of physiological arousal and cognitive functioning is curvilinear, and that optimal cognitive functioning is only possible at intermediate levels of arousal. From this perspective, emotions that come into play in a situation of conflict are far too intense to allow for cognitive functioning on a level anywhere close to optimal.

Secondly, a demand for crucial decisions to be made quickly and correctly, combined with an information overload, is one of the mechanisms leading to psychological stress, which also reduces the individuals' information processing complexity (Schroder et al. 1967; Suedfeld and Tetlock 1977). This reduction results in less search for new information, less accurate distinctions between things that are relevant or irrelevant, suppressing unwanted input, a tendency to achieve premature closure, in-group conformity, stereotyped attitudes and responses, only robust distinctions between pieces of information, as well as quicker decisions (Suedfeld and Tetlock 1977).

A study concerning integrative complexity of information processing of the Kennedy administration before, during and after the Cuban missile crisis in October 1962 (Guttieri et al. 1995) confirms that emotional tension related to conflict has a negative effect on the cognitive capacities of its participants. In this study, changes in integrative complexity of written and oral statements, as well as private materials of Kennedy's close advisors, were charted. The authors show an increase in complexity of the politicians when a challenge was first recognized and coping mechanisms were activated (pre-crisis to early crisis period), and a decrease when the conflict escalated and continued without resolution (early to late crisis period). The lowest indices of integrative complexity were noted on October 27th, when an American reconnaissance plane was shot down over Cuba, and Kennedy received a letter from Khrushchev suggesting removal of U.S. missiles from Turkey. That day, the threat of hot war was the greatest.

When one examines the relationship between conflict and its participants' cognitive complexity more closely, it becomes even more complicated. A study conducted by Driver (1962, as cited in Raphael 1982), in which he employed Guetzkow's InterNation Simulation (1959) – an interactive exercise in which the participants act as country representatives and take foreign policy decisions – shows that complexity scores may depend curvilinearly upon the characteristics of the environment. In simulation runs classified as imposing little or no stress, the level of integrative complexity exhibited by the participants was the lowest. In runs involving moderate levels of stress, i.e., peacefully resolved conflict, the participants exhibited highest levels of integrative complexity, while drastically less complex information processing was exhibited in runs that involved war.

Cognitive complexity is related to the attitudes that people adopt in a situation of conflict. When a group of politicians was faced with a direct, emotionally laden attack on their positions in a real-life conflict, those with lower cognitive skills adopted more competitive attitudes, while those with higher cognitive complexity sought more cooperative and avoidance solutions (Golec 2002a, b). An archival analysis of data from seven twentieth-century conflicts has shown that cognitive complexity may even affect the course taken in a given conflict (Suedfeld and Tetlock 1977). Diplomatic communication in international crises that ended in war was compared with that in crises that settled peacefully. It turned out that the complexity of messages produced by political leaders in crises that resulted in war was significantly lower.

Additionally, as a conflict approached its climax, the complexity of communication declined when it was leading towards war, and increased when it headed towards a peaceful settlement (Suedfeld and Tetlock 1977). These results were also confirmed in a study of communication during the Middle East conflict, between 1947 and 1976 (Suedfeld et al. 1977). The results show that complexity of information processing was significantly reduced in UN General Assembly speeches made in the months preceding the outbreak of war. Decreases in cognitive complexity taking place prior to war, the onset of violence, or other uncompromising, uni-dimensional crisis solutions, have been explained in terms of disruptive stress to which the decision makers are exposed (Suedfeld and Bluck 1988; Suedfeld et al. 1977). Important problems receive a major share of attention and resources, which leads to high complexity of information processing. However, no resources are endless, so if the situation continues too long without resolution, becomes threatening, or time pressure arises, complexity of the decision maker drops significantly (Guttieri et al. 1995). A decrease in the participants' cognitive complexity can even be treated as a predictor of an approaching escalation of conflict (Raphael 1982).

A simple model capturing the relationships between conflict and cognitive complexity can be inferred from the results presented above: (a) emotions evoked by a situation of conflict impair the participants' cognitive functioning; (b) environmental and cognitive complexity are curvilinearly related – as the complexity of the environment increases, so does people's cognitive complexity, until a threshold reversing this relationship is reached; (c) complexity levels are related to attitudes and outcomes of conflict in such way that more complex information processing leads to more cooperative, as opposed to competitive, attitudes and in consequence to more peaceful, as opposed to violent, outcomes.

Individual Differences in Cognitive Complexity and Conflict

Possessing advanced cognitive skills does not mean that one will take full advantage of them, especially in situations that are important and demanding. The extent to which individuals make use of their cognitive skills is determined by many factors, such as emotion, stress, or motivation. Conflict involves some of these factors, and therefore has the power to simplify cognitive processing, compared to the participants' full potential. However, not only do individuals vary in their attained levels of cognitive complexity, but also the effect that being involved in conflict exerts on cognitive functioning is not universal within these levels (Golec 2002a, b).

Individual differences in cognitive skills, their relationship to attitudes in conflict, and how these are influenced when faced with a direct attack on one's positions, were the focus of a quasi-experimental study conducted on Polish politicians (Golec 2002a, b). The study was designed around a real-life political conflict, which took place in Poland in 1997 – a conflict about the ratification of a concordat. In the study, the politicians who were actually parties to the conflict were interviewed, to assess their level of cognitive development (Rosenberg 1988), and divided into groups of 'simple' and 'complex' thinkers. Simple thinkers were those who have only achieved Rosenberg's linear stage, and thus were unable to de-center, while complex thinkers were at least able to transcend their own perspective. After the politicians had been assigned to the two groups, half of each group was presented individually with an excerpt of a mock newspaper article about the conflict (attack condition), and then asked questions concerning their attitudes towards the conflict, the answers to which were coded as competitive, cooperative or avoidance-oriented. The other half of each group also answered attitude questions, but did not read any article (neutral condition).

The results show that the approaches toward conflict do indeed vary according to attained cognitive skills. In the neutral condition, where participants were hypothesized to be able to make use of all of their cognitive potential, the majority of simple thinkers adopted competitive attitudes, while the majority of complex thinkers adopted cooperative ones. In the emotional attack condition, simple thinkers held on to their competitive attitudes, while complex thinkers chose to avoid conflict. Across conditions, simple thinkers were more likely to express competitive attitudes, hold negative images of the opponents and to blame them for the situation, aiming at confrontation and the use of force. Complex thinkers, on the other hand, avoided competition, wanted communication and peaceful resolution, remained neutral in their evaluations of the opponent, and blamed no one in particular.

It thus seems that in a situation of conflict, the level of cognitive development differentiates individuals in two ways. On the one hand, at the simpler end of the continuum, a tendency to adopt more competitive strategies is stronger than at the complex end, which is characterized by more cooperative and avoidance attitudes. On the other hand, the confrontational inclination of the simple thinkers is independent of the characteristics of the situation, while the attitudes adopted by more complex thinkers appear more flexible, suited to the situational context.

Conflict as a Collapse of Complexity

Dynamic systems theory (or complexity science) is an increasingly influential paradigm in various areas of science (Myers 2009; Solomon et al. 2009; Vallacher et al. 2002) that also offers innovative ideas for conceptualizing and addressing conflict. From this perspective, one on which this section is based, conflict is characterized by a set of elements, such as specific beliefs, actions or feelings, which are all linked to one another (Coleman et al. 2006, 2007; Nowak et al. 2006). Changes in any element, e.g. the level of violence, are influenced by, and in turn influence, changes in other elements, such as hostile attitudes, stereotypical perceptions, or trust between the parties.

Considering conflict in terms of dynamical systems has three major implications for its analysis: (a) conflicts being inherently dynamic escalate and deescalate, spread into new areas, and can be passed from generation to generation. Elements constituting the system have their own internal dynamics, so conflicts can develop without any external influences; (b) any attempts at intervention do not induce direct change, but only perturb the system's internal dynamics. Conflict may respond to an external intervention by resisting it, responding in an exaggerated manner, evolving in a completely unpredictable direction, or behaving according to the intervener's plan; (c) the dynamic nature of conflict makes it impossible to predict what the specific outcomes of any intervention will be. However, dynamic systems show general patterns of dynamics, which are stable and predictable on a longer time scale. If a conflict persists, it is likely to develop patterns of dynamics, which define the relationships between psychological and social mechanisms operating within and between individuals and groups. Once the parties have developed stable patterns of thought and action towards one another, the conflict gets detached from the issues that initiated it, and functions as a larger system with dynamic properties.

In healthy relationships, conflicts are usually confined to specific issues, leaving many issues about which no conflict exists. In such cases, conflicts can be solved constructively, promoting relationship maintenance and growth. The mechanisms operating on different issues may even balance each other, so when a conflict arises in one domain, it fosters positive responses in others, in order to maintain the overall relationship. For instance, when a conflict about spending a weekend springs up in a couple, the threat that it poses for the relationship may be compensated by extra care and attentiveness between the partners in other domains. Therefore, healthy relationships (on both interpersonal, as well as inter-group level) are complex and multidimensional, with various mechanisms operating independently with respect to different issues, often in a compensatory manner.

When a conflict escalates, the complexity of the system collapses. Features that were normally independent, or functioned in opposition, become aligned and work in a mutually reinforcing manner. This collapse of multidimensionality has two aspects: (a) the appearance of positive feedback loops between previously independent elements, i.e. when the activation of one element increases the activation of other elements, to which it is linked, and (b) the disappearance of negative feedback loops between elements, i.e. when the activation of an element decreases the activation of others. Loss of balance between positive and negative feedback loops fuels the potential for destructive, rather than constructive, conflict. In a healthy relationship, signs of damage to the other party tend to halt further attacks. But when multidimensionality is reduced, and negative feedback loops have reversed into positive ones, "the tears of our enemy augment, rather than inhibit, aggression," (Coleman et al. 2007) which can only lead to conflict stabilization or further escalation.

Conflict is manifested at different levels of social reality. At the individual level, it is reflected in thoughts, feelings and actions of a single person; at the interpersonal level, it influences the dynamics of interpersonal relations; at the inter-group level, it is visible in the interactions between social groups. Destructive conflict not only links elements belonging to the same level, but also introduces links between levels, so that mechanisms operating at one level promote conflict at others. It means that conflict launched at the inter-group level is likely to influence interpersonal emotions and attitudes, so even when the original issue has been resolved, the links to other levels may reinstate the conflict. Also, the behavior of an individual may lead to the escalation of conflict on an inter-group level. Through social interaction, a shared reality of group members emerges, so even individuals that were never directly involved, develop the sense of conflict through the group's shared reality.

The alignment of previously independent conflict elements reflects a drive for coherence in psychological and social systems (see Festinger 1957; Heider 1958), a common feature of emotions (e.g. Thagard and Nerb 2002), self-concept (Nowak, Vallacher and Strawińska, Chap. 2), or social judgment (e.g. Vallacher et al. 1994). A variety of factors can reinforce this tendency and promote the appearance of linkages between different issues. Personal experience of a co-occurrence of factors links these elements in such way that whenever one of them appears, all others are automatically activated. If a given situation of conflict over common resources repeatedly escalates into harsh words, anger and hostile attitudes, these elements are likely to become bound into a single structure so that any one element activates the entire ensemble. Moreover, the presence of intense emotions promotes a heightened pressure for coherence. Under the influence of emotions, the view of the other is simplified. It is then difficult to detect the nuances in a given situation, or appreciate the complexity of issues. Strong emotions can also intensify positive feedback loops between levels, by promoting stereotypical thinking about the whole social group to which the other party belongs. As a consequence, one party develops a global and emotionally loaded attitude toward the other, which no longer allows for subtle differentiation between the issues in conflict, and those outside it.

When complexity collapses, conflict intensity is transformed into an essentially binary variable – there is no room for its gradual escalation, it is either absent or present. Any single conflict related element activates all related features, so the reaction to one is the same, as it would be to all. The diversity of possible responses, as well as the flexibility of the system is lost. Lack of negative feedback loops makes any action, once initiated, hard to stop, since all connections to other elements only reinforce it. This alignment of different conflict elements makes conflicts extremely difficult to de-escalate. Resolving the issue that first initiated it no longer does the job, since the remaining elements continue to operate and fuel the conflict. For a de-escalation attempt to be successful, not only the issue that first initiated a conflict, but also all the others, to which it has become connected, must be dealt with.

From this dynamic perspective, attempts at conflict resolution should focus on restoring the system's multidimensionality. There are no universal guidelines; every conflict is different and requires a careful case study. First, the relevant elements and links between them must be identified, and then the linkages between key elements should be disturbed or, when possible, broken. This allows for the issues to be decoupled and addressed separately. For example, if in the course of the conflict a generalized negative attitude toward the out-group has developed, showing individual positive examples of out-group members that do not conform to the stereotype should disturb the homogenized perception. Under certain circumstances (Allport 1954; see the review in Amichai-Hamburger and McKenna 2006), contact between individual members of the conflicting groups should increase the complexity of the perception of out-group members. Alternatively, one could try to destabilize the borders between the in-group and out-group by strengthening identities not related to conflict, such as professional roles or common interests, rather than ethnic or national ones. In the conflict over democracy in communist Poland in the 1980s, the Round Table negotiations, where communist leaders talked with delegates of the opposition, were essentially a restoration of the systems' complexity. A uni-dimensional struggle for power was transformed into a number of small, issue-specific conflicts – such as education, the media or healthcare – which were then dealt with independently. Restoring the multidimensionality of the system is by no means an easy task, and it gives no guarantee for a successful conflict resolution, but it does make the grounds for addressing each of the conflict elements in separation.

Summary and Conclusions

It is now time to go back to the story from the beginning of this chapter, and try to explain it in terms of cognitive complexity. The relationship between my climbing partner and myself was healthy and multidimensional. When the disagreement broke out, it escalated fast, facilitated by fatigue, emotions, and additional stress evoked by the storm. The multidimensionality of our perceptions of each other collapsed, so there was no more room for ambiguity – the perception of the other had to be either positive or negative on every possible dimension. This also led to more competitive attitudes and the awakening of a fighting spirit. From my partner's perspective, a correlation between my perceived cartography skills, personality traits and physical attributes appeared. He could no longer bear the situation in which I was to some extent 'bad' and 'good' at the same time, not even good-looking. He wanted to take me down quickly, so he used an argument from the domain of physical appearance, which was the first association with me that came to his mind. What seemed like a silly conflict over the best route back to the camp changed his standards of attractiveness, it impaired his ability to perceive our relationship on a number of independent dimensions, even if only temporarily.

Only a brief overview of the theories and research related to cognitive complexity and conflict was presented in this chapter. But even this small sample makes it quite clear that complexity constitutes an important parameter of social relations, and plays an important role in the development, escalation and resolution of conflicts. Cognitive complexity is only one of its many facets, the most salient one from a psychological perspective. It thus seems that the complexity science approach provides a promising framework for further studies of conflict, and may serve as an integrative platform between different theories and levels of analysis.

References

Adorno, T.W., Frenkel-Brunswik, E., Levinson, D.J., Sanford, R.N.: The Authoritarian Personality. Harper and Row, New York (1950)

Allport, G.: The Nature of Prejudice. Addison-Wesley, Reading (1954)

Amichai-Hamburger, Y., McKenna, K.Y.A.: The contact hypothesis reconsidered: interacting via the Internet. J. Comput. Mediat. Commun. 11(3), article 7 (2006), http://jcmc.indiana.edu/vol11/issue3/amichai-hamburger.html. Retrieved 15 Apr 2011

Baker-Brown, G., Ballard, E.J., Bluck, S., de Vries, B., Suedfeld, P., Tetlock, P.E.: The conceptual/integrative complexity scoring manual. In: Smith, C.P., Atkinson, J.W., McClelland, D.C., Veroff, J. (eds.) Motivation and Personality: Handbook of Thematic Content Analysis, pp. 393–418. Cambridge University Press, New York (1992)

Bieri, J.: Cognitive complexity-simplicity and predictive behavior. J. Abnorm. Soc. Psychol. 51, 263–268 (1955)

Bieri, J., Atkins, A.L., Briar, S., Leaman, R.L., Miller, H., Tripodi, T.: Clinical and Social Judgement: The Discrimination of Behavioral Information. Wiley, New York (1966)

Chandler, M.J.: Adolescence, egocentrism, and epistemological loneliness. In B.Z. Presseisen, D. Goldstein, M.H. Appel (Eds.) Topics in Cognitive development, vol. 2. Plenum, New York: (1978)

Coleman, P.T., Bui-Wrzosinska, L., Vallacher, R., Nowak, A.: Protracted conflicts as dynamical systems: guidelines and methods for intervention. In: Schneider, A., Honeyman, C. (eds.) The Negotiator's Fieldbook, pp. 61–74. American Bar Association Book, Chicago (2006)

Coleman, P.T., Vallacher, R., Nowak, A., Bui-Wrzosińska, L.: Intractable conflict as an attractor: presenting a dynamical model of conflict, escalation, and intractability. Am. Behav. Sci. 50(11), 1454–1475 (2007)

Crain, W.C.: Theories of Development. Prentice-Hall, Upper Saddle River (1985)

Crockett, W.H.: Cognitive complexity and impression formation. In: Maher, B.A. (ed.) Progress in Experimental Personality Research, vol. 2, pp. 47–90. Academic, New York (1965)

Deutsch, M.: The Resolution of Conflict: Constructive and Destructive Processes. Yale University Press, New Haven (1973)

Festinger, L.: A Theory of Cognitive Dissonance. Row, Peterson, Evanston (1957)

Golec, A.: Cognitive skills as predictor of attitudes toward political conflict: a study of Polish politicians. Polit. Psychol. 23(4), 731–759 (2002a)

Golec, A.: Konflikt polityczny. Myślenie i emocje. Wydawnictwo Akademickie Dialog, Warszawa (2002b)

Guetzkow, H.: A use of simulation in the study of inter-nation relations. Behav. Sci. 4, 183–191 (1959)

Guttieri, K., Wallace, M.D., Suedfeld, P.: The integrative complexity of American decision makers in the Cuban missile crisis. J. Confl. Resolut. 39, 595–621 (1995)

Heider, F.: The Psychology of Interpersonal Relations. John Wiley, New York (1958)

Hermann, M.G.: Assessing the personalities of Soviet Politburo members. Pers. Soc. Psychol. Bull. **6**, 332–352 (1980)

Kelly, G.A.: The Psychology of Personal Constructs. Norton, New York (1955)

Mitchell, T.R.: Leader complexity and leadership style. J. Pers. Soc. Psychol, **16**, 166–174 (1970)

Myers, R.: Encyclopedia of Complexity and Systems Science. Springer, Berlin (2009)

Nowak, A., Vallacher, R., Bui-Wrzosińska, L., Coleman, P.T.: Attracted to conflict: a dynamical perspective on malignant social relations. In: Golec, A., Skarżynska, K. (eds.) Understanding Social Change: Political Psychology in Poland. Nova Science Publishers Ltd., Hauppauge (2006)

O'Keefe, D.J., Shepherd, G.J., Streeter, T.: Role category questionnaire measures of cognitive complexity: reliability and comparability of alternative forms. Cent. States Speech J. **33**, 333–338 (1982)

Piaget, J.: Six études de psychologie. Edition Gonthier, Genève (1964)

Raphael, T.D.: Integrative complexity theory and forecasting international crises: Berlin 1946–1962. J. Confl. Resolut. **26**(3), 423–450 (1982)

Rosenberg, S.W.: Reason, Ideology and Politics. Princeton University Press, Princeton (1988)

Schroder, H.M., Driver, M.J., Streufert, S.: Human Information Processing. Holt, Rinehart & Winston, New York (1967)

Scott, W.A.: Cognitive complexity and cognitive flexibility. Sociometry **25**, 405–414 (1962)

Scott, W.A., Osgood, D.W., Peterson, C.: Cognitive Structure: Theory and Measurement of Individual Differences. Winston, New York (1979)

Seaman, J.M., Koenig, F.: A comparison of measures of cognitive complexity. Sociometry **37**, 375–391 (1974)

Sidanius, J.: Cognitive functioning and socio-political ideology: an explorative study. Percept. Mot. Skills **46**, 515–530 (1978)

Solomon, S., Bottazzi, G., Brée, D.S., Cantono, S., Louzoun, Y., Shnerb, N.M., Nowak, A., Vignes, A., Weisbuch, G.: Common Complex Collective Phenomena: Implications for Economic and Social Policy Making. E. Bendyk and A. Zdrodowska, Warsaw (2009)

Suedfeld, P.: Indices of world tension in The Bulletin of the Atomic Scientists. Polit. Psychol. **2**(3/4), 114–123 (1980)

Suedfeld, P., Bluck, S.: Changes in integrative complexity prior to surprise attacks. J. Confl. Resolut. **32**, 626–635 (1988)

Suedfeld, P., Tetlock, P.: Integrative complexity of communications in international crises. J. Confl. Resolut. **21**, 169–184 (1977)

Suedfeld, P., Tetlock, P.E., Ramirez, C.: War, peace, and integrative complexity. J. Confl. Resolut. **21**, 427–442 (1977)

Suedfeld, P., Tetlock, P.E., Streufert, S.: Conceptual/integrative complexity. In: Smith, C.P., Atkinson, J.W., McClelland, D.C., Veroff, J. (eds.) Motivation and Personality: Handbook of Thematic Content Analysis, pp. 393–400. Cambridge University Press, New York (1992)

Thagard, P., Nerb, J.: Emotional gestalts: appraisal, change, and the dynamics of affect. Pers. Soc. Psychol. Rev. **6**, 274–282 (2002)

Vallacher, R.R., Nowak, A., Kaufman, J.: Intrinsic dynamics of social judgment. J. Pers. Soc. Psychol. **66**, 20–34 (1994)

Vallacher, R.R., Read, S.J., Nowak, A.: Special issue: the dynamical perspective in personality and social psychology. Pers. Soc. Psychol. Rev. **6**(4), 264–388 (2002)

Van Hiel, A., Mervielde, I.: The measurement of cognitive complexity and its relationship with political extremism. Polit. Psychol. **24**(4), 781–801 (2003)

Vannoy, J.S.: Generality of cognitive complexity-simplicity as a personality construct. J. Pers. Soc. Psychol. **2**, 385–396 (1965)

Zimbardo, P.G.: Psychology and Life, 13th edn. Harper Collins, New York (1992)

Chapter 7
Social Entrepreneurs Open Closed Worlds: The Transformative Influence of Weak Ties

Ryszard Praszkier

Abstract The concept of weak versus strong social ties is introduced, including the classical and contemporary definitions as well as the ambiguities related to the operationalization of the definitions. Furthermore, a concept is presented of social entrepreneurship and the way social entrepreneurs build and enhance weak ties in disenfranchised groups and communities. Analogies to chemical processes (using a static as well as a dynamic model) provide a gateway for further research and for modeling the dynamics that measure the strength of social ties. One of the conclusions is that for a harmonious development of groups, communities and societies, a balance between strong and weak ties should be sustained.

Introduction

One area of inquiry that has intrigued social scientists is exploring the dynamics of how societies change. There are of course many ways: some changes come along with technological development; for example, ICT[1] connected people around the globe, including those in remote and isolated areas. It can also be triggered by economic stimuli, when for example new investments provide jobs and growth opportunities. Change in local communities may also occur because the entire country is going through political and economic transformation (for example, the Balkans in the 1990s).

These are examples illustrating exogenous change, that is, change triggered from outside the family, group or community. However, there also exists a different kind of change, endogenous change (Noble 2000), that is, coming from the inside. The following is an illustration:

---

[1] ICT is Information and Computer Technologies.

R. Praszkier
Institute for Social Studies, University of Warsaw, Stawki 5/7, 00-183 Warszawa, Poland
e-mail: ryszardpr@gmail.com

A. Nowak et al. (eds.), *Complex Human Dynamics*, Understanding Complex Systems, 111
DOI 10.1007/978-3-642-31436-0_7, © Springer-Verlag Berlin Heidelberg 2013

In an economically and socially disadvantaged region of southern Poland (Żegocina), several top-down attempts at addressing the society's plight had failed, evoking frustration and resistance to outside experts who had little knowledge of the population. For example, numerous experts offered top-down solutions based on templates and macro-analyses, developed in the absence of a basic understanding of the history, needs, frustrations and latent capabilities of this particular community. Not surprisingly, their costly efforts to address the society's difficulties failed, resulting in growing frustration and anger and engendering among the inhabitants a sense of powerlessness and even lower self-esteem, as well as strong feelings of distrust and aggression.

Ashoka Fellow Dagmara Bienkowska,[2] a young, passionate university alumna and social activist, offered a ray of hope for this beleaguered population. She started her intervention by learning about the community from within, living there, talking to people and learning about their suffering and their dreams. Soon she understood that despite their hard existence there lay a strong identification with the region and dreams for its development.

Analyzing the sociological structure, she noticed that there were two marginalized groups: senior citizens and aggressive, troublesome young people. She understood, on the other hand, that both groups played significant roles in the community: the older adults exercised a strong behind-the-scenes influence on others, whereas the problematic youth occupied a prominent place on the list of negative stories that traveled through the community's gossip mill and acted as a target for shared frustrations.

Dagmara's thought was to connect those two completely disparate groups, believing that together they might become a source of positive energy. Her plan was to devise a joint undertaking: she suggested that the young people visit the senior citizens and gather some recipes of regional dishes. This worked out perfectly, as they were more than eager to share their traditional recipes, and the young people felt that they were doing something new and important. The astonishing image of local bullies and senior citizens working on a project together got the attention of the local authorities, who saw this as an opportunity and proceeded to print an unedited edition of the Cookbook of Żegocina County. The book was distributed at conferences as a local product of which they could be proud; a second edition was published professionally, sold out, and the income from it channeled into community educational projects.

The success of the project transformed the youth group into a major entrepreneurial force, as they saw that cooperation yielded an immense payback. They launched several new ventures, triggering an entrepreneurial movement among other community members. In a few years, Żegocina had the largest number of newly established business ventures in the region.

This transformation strengthened the people of Żegocina in a variety of unforeseen ways. For instance, during a time of national disaster, when the region was

---

[2] See: www.ashoka.org/fellows/dagmara_bienkowska; more on Ashoka in the next section.

heavily affected by floods, it turned out that this community was the best organized and the best equipped to cope with the disaster. Capitalizing on this development, a Żegocina Flood Book was published and sold as a manual for handling natural disasters. Once again the income was channeled into the community's social projects. Eventually, through their bottom-up approach, Żegocina County experienced unanticipated economic development, based on their own potential and surpassing all their neighboring communities.

This narrative leads to the following conclusions: the bottom-up development based on a group's own, though sometimes latent, capabilities may become a vehicle for durable and sustainable change. In this case, the endogenous process was triggered by a connector (Gladwell 2002), as Dagmara Bienkowska's intervention was based solely on connecting two isolated, though strongly self-interconnected groups. This would indicate that connecting distant and isolated groups may activate an endogenous change process (Csermely 2008); in other words, linking the two groups served as a catalyst. Moreover, after Dagmara's intervention, the entire network consisted of strongly internally connected groups, as it did before, though in between those previously isolated groups some new connections appeared, which led to joint initiatives, and which resulted in the emergence of community-bonding social capital (Putnam 2000).

This article will focus on the significance of the strength of social ties in the process of social transformation. The existing literature mainly focuses on a static analysis, whereas this chapter explores the behavioral dynamics of the strength or weakness of ties during the transformation process. It will concentrate particularly on the dynamics of the change launched by social entrepreneurs, passionate individuals who are known for bringing new ideas to address insurmountable social problems and for merging creativity with an entrepreneurial approach, and whose ideas, with minimal initial investments, usually rise to the national or international level. The best example is Mohammad Yunus, the winner of the 2006 Nobel Peace Prize: starting in Bangladesh, one of the poorest countries in the world, Mohammad Yunus and his Grameen Bank launched the idea of 'banking for the poor.' The concept was to offer microcredits to poor women in the form of revolving loans, which enabled them to launch their own small-business ventures. This idea proved successful for thousands of Bangladeshi families and spread globally, changing the lives of millions through the microfinance system, empowering the poorest of the poor.

This chapter posits that the social transformation introduced by social entrepreneurs has an important component, often not visible and not knowingly intended: that is, regulating the strength of the social ties within a group, thereby maintaining the balance between strong and weak ties. In some cases, it is accomplished by opening windows to the outside world, so that previously isolated individuals, groups or societies become connected with others through weak ties; in other cases, it is accomplished by increasing the number of strong ties, so as to restore individual, group, or societal identity and strength emanating from deep cultural roots.

The premises for this chapter are two: one relates to the concept of 'The Power of Weak Ties,' a phrase coined by Mark Granovetter (1973) which is gaining in currency in the academic world; the other refers to the theory and practice of social entrepreneurship and the unique way social entrepreneurs introduce durable, irreversible and lasting social change (Bornstein 2004; Bornstein and Davis 2010; Elkington and Hartigan 2008).

The first section will introduce the phenomenon of social entrepreneurship based on two examples; the second section will explore the concept of weak ties; and the third will focus on the psycho-sociological aspects related to establishing both strong and weak ties. The fourth section will delineate the process of societal transformation in relation to the strength or weakness of ties, and the fifth will illustrate how social entrepreneurs spark and facilitate the change process so that one of its emergent results is the propensity for establishing weak ties and, as a consequence, opening up the field of possibilities for multiple interconnections with the outer world. The final section will present a dynamic model of the strength or weakness of ties of societies in transition, as an analogue to the phase changes of water molecules.

Social Entrepreneurship

Social entrepreneurs usually address seemingly unsolvable social problems, and in so doing, generate a huge impact on the social landscape. This is often done by triggering a bottom-up process involving and empowering society as a whole. Peter Drucker framed the concept by noting that social entrepreneurs change the performance capacity of society (Gendron 1966), meaning that the impact of social entrepreneurs exceeds by far their specific areas of interest by empowering societies to enhance their overall performance.

There is an increasing interest in the field of social entrepreneurship among academics and social activists (Gentile 2002; Leadbeater 1997; Steyaert and Hjorth 2006) as well as among many in the private sector (Anonymous Busines Editors 2002; Brinckerhoff 2000; Gentile 2002). According to Mair et al. (2006), social entrepreneurship has made a popular name for itself globally as a 'new phenomenon' that is reshaping the way we think about social-value creation.

The Classical Definitions

One of the classical definitions of social entrepreneurship and the social entrepreneur is provided by Dees (1998), who says that social entrepreneurs play the role of change agents in the social sector by:

- adopting a mission to create and sustain a social value (not just private value);
- recognizing and relentlessly pursuing new opportunities to serve that mission;
- engaging in a process of continuous innovation, adaptation, and learning;
- acting boldly without being limited by resources currently at hand;
- exhibiting a heightened sense of accountability to the constituencies served and for the outcomes created.

Bornstein (2004) considers the Ashoka definition of social entrepreneurship the most comprehensive. Ashoka, Innovators for the Public,[3] is an international association, operating since 1980 in over 70 countries, whose mission is to empower social entrepreneurs. According to Ashoka (2000), social entrepreneurship can produce small changes in the short term that reverberate through existing systems, ultimately effecting significant change in the longer term. The selection criteria for Ashoka Fellows constitute the definition of social entrepreneurship; according to Drayton (2002, 2005) and Hammonds (2005), they are:

- having a new idea for solving a critical social problem;
- being creative;
- having an entrepreneurial personality;
- envisioning the broad social impact of the idea;
- possessing an unquestionable ethical fiber.[4]

The Dynamical Delineation

Praszkier and Nowak (2011) introduce a dynamical understanding of social entrepreneurship, to wit: social entrepreneurs initiate and manage the process of emergence. Nowak (2004) holds that the basic idea of emergence is that the local interactions among low-level elements, where each element adjusts to other elements without reference to a global pattern, may lead to the emergence of highly coherent structures and behaviors on the level of the whole. Social entrepreneurs achieve that through creating an enabling environment for initiating bottom-up social networks, as well as by providing the opportunity for social coordination, so that the networks operate in a cohesive way, leading to emergent occurrences.

Praszkier and Nowak (2011) assert that the way social entrepreneurs create enabling structures is mainly through social networks that form around their ideas. Furthermore they have concluded that, under the influence of the social entrepreneur, previously isolated groups become open to establishing **weak ties**, which connect those groups with the – much bigger and most vibrant and resilient – **scale-free** network. Such a network creates a propitious environment for the appearance of emergent social phenomena, which become a solid part of the

---

[3] See: www.ashoka.org.

[4] See: Ashoka selection criteria explained: www.ashoka.org/support/criteria.

society; as an additional effect, it prompts a problem-solving culture, indispensable for tackling possible future problems.

The Strength or Weakness of Ties

History

The power of weak ties was proposed in the late 1960s by Mark Granovetter when conducting research for his doctoral thesis on the subject of how people 'network'; one of the dimensions explores the ways people use their connections in order to find a job (Barabási 2003). The respondents were asked such simple questions as if it was a friend who helped them find their current job. The predominant response was: no, not a friend, an acquaintance – a weak connection, to be sure – and one that reminded Mark Granovetter of something he had learned as a student of chemistry: that hydrogen bonds, weak in and of themselves, nevertheless are responsible for holding huge water molecules together. And thus the concept of the Strength of Weak Ties (Granovetter 1973) came into being.

Definition

For the purposes of this chapter, we are following the lead of Granovetter (1973), who, in assessing the strength of the ties, readily admits that tie strength is an intuitive judgment; he characterizes the weak tie as a combination (probably linear) of four indicators: how long the tie has existed, the emotional intensity, the intimacy, and the reciprocal services. He also acknowledges that each of these qualities is somewhat independent of the other, and yet they are at the same time correlated.[5] It is also important to point out that the considerable ambiguity and inconsistency in this definition fails to fully elucidate how the four indicators combine to create tie strength; in other words what 'weight' do we assign them and does each count equally? At what point does a tie move from weak to strong? (Krackhardt 1992). However, despite these questions, this intuitive set of four dimensions has been applied for over three decades by several researchers, for example Petróczi et al. (2007).

It was Peter Csermely (2009) who offered a new definition: a link is defined as weak when its addition or removal does not change the mean value of a target measure in a statistically discernible way. And, again, a caveat: "I am aware that, like all functional definitions, this one is also highly context-dependent. For this

---

[5] Granovetter (1973) held that for his article it is sufficient to estimate roughly, on an intuitive basis, whether a given tie is strong, weak, or absent.

definition, we have to set a target measure, we have to be able to add or remove the link, we have to be able to repeat the determination of the measure several times and, the most difficult condition, we have to maintain all conditions of the network intact (apart from the addition or removal of the link) between these measurements" (Csermely 2009, p. 107).

Both Granovetter and Csermely have put forward a relatively precise general definition, and both admit its intuitive operationalization, leaving room for suiting the operational implications to the research context.

Implications

There is a growing understanding of the significance of how weak and strong ties work among individuals and groups in relation to the outside world, especially with regard to access to information and resources or to mobility. A weak tie between an individual and his acquaintances becomes a crucial bridge between two densely knit clumps of close friends. The significance of weak ties is that they are far more likely to be bridges between even distant network participants or groups, than are strong ties (Granovetter 1973, 1983).

A Few Weak Ties

Individuals whose social connections consist of strong ties plus only a few weak ties will be deprived of information from distant parts of the social system and will be confined to the provincial news and views of their close friends. This deprivation will not only insulate them from the latest ideas and trends but may put them at a disadvantage in the labor market, where time is of the essence when it comes to seizing opportunities for advancement. Moreover, they will be poorly integrated into political or other goal-oriented movements, since membership typically results from being recruited by friends (Granovetter 1973, 1983, 1995). Social systems lacking in weak ties will be fragmented and incoherent (Granovetter 1983).

Strong ties are relationships among people who work, live or play together. These connections are utilized frequently and need considerable management to stay healthy. People with strong ties, and the frequent interactions they engender, tend to think alike over time. This phenomenon tends to reduce the diversity of ideas, and in worst-case scenarios, lead to 'groupthink' (Porter 2007).

Many Weak Ties

Establishing weak ties requires cognitive flexibility, and an ability to function in complex voluntary organizations. The weaker the ties the more likely the individual will have access to heterogeneous resources; weak ties have a special role in a person's opportunity for mobility (Granovetter 1973, 1995; Lin 2001).

The study confirmed that weak ties were the ones that resulted in new jobs. Moreover, those whose jobs were found through strong ties were far more likely to have had a period of unemployment between jobs than those using weak ties (Granovetter 1983, 1995).

Weak ties provide communication opportunities between members of different groups. They are utilized infrequently and therefore don't need much management to stay healthy. They lead to a diversity of ideas, as they tie together disparate modes of thought (Porter 2007).

Example: Family Dynamics

Obviously no family unit escapes conflicts and crises. Nuclear families with a small number of weak ties in a crisis situation are dependent solely on strong bonds. For example, young parents in conflict over child-rearing methods may seek their parents' support, thus limiting the sources of information or advice to the existing close-knit circles. It may happen that the grandparents, set in their own, not necessarily complementary beliefs, end up deepening the conflict between the parents. This process, in a feedback loop, may lead to the further escalation of conflict and to the eventual dissolution of the nuclear family. However, the availability of many weak bonds opens the door to allow the influx of a greater variety of information and sharing of experience, which may also include advice on cooling the conflict.

Analogies to Chemical Bonds

In the analysis of the results of his research on job-hunting Granovetter (1983) drew an analogy with the way 'weak' bonds between hydrogen atoms hold huge water molecules together; those huge molecules are themselves held together by 'strong' bonds (see Fig. 7.1).

A similar combination of strong and weak bonds, according to Granovetter, holds the members of society together, see Fig. 7.2.

Sociopsychological Aspects of Establishing Strong or Weak Ties

We know, intuitively, that it takes certain personality types to take on the task of establishing ties, be they strong or weak.

Granovetter asserted that weak ties are exactly the kind that lead to complex role sets and the need for **cognitive flexibility**. Furthermore, he states "the ability to

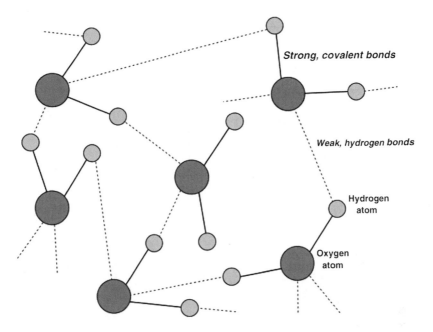

Fig. 7.1 When water is in the liquid state, freely floating water molecules (one oxygen atom bonded with two hydrogen atoms, *solid lines*) connect with each other through weak 'hydrogen bonds' (*dashed line*)

function in complex interrelations may depend on a habit of mind that permits one to assess the needs, motives, and actions of a great variety of different people simultaneously" (Granovetter 1983, p. 205). The latter would mean a special capability to maintain a **diffused empathy**. Also Granovetter (1983) holds that weak ties have a special role in a person's **propensity for mobility**.

Csermely (2009) makes a distinction between those who are prone to establishing mainly strong ties (calling them 'stronglinkers') and those who are prone to establishing weak ties ('weaklinkers').[6]

Stronglinkers and weaklinkers are equally important: the first are necessary to build the core of our networks and, even more importantly, they are the only solution during hard, stressful times. Weaklinkers are useful in periods of expansion; however, they have to be controlled by stronglinkers, otherwise the network will overspend its resources, and become inefficient, overconnected and unstable.

Csermely (2009) points out several personality differences between the stronglinker and weaklinker:

---

[6] Csermely goes even further, indicating that there are fairly distinct stronglinker and weaklinker phenotypes.

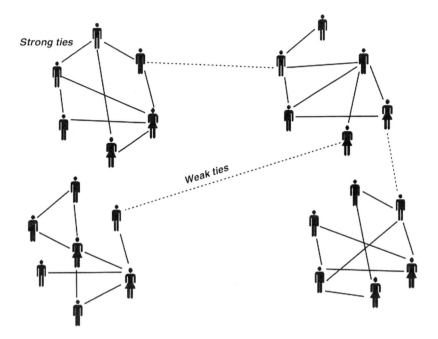

Fig. 7.2 Small groups of strong ties are interrelated with other groups through a few weak ties. Everybody from all close knit groups is in this way connected with everybody else in the entire network

- **Friendship networks**: Stronglinkers tend to rely on family links, usually having a few good friends, and they prefer and excel at creating reliable, lifelong contacts; weaklinkers, on the other hand, have friends from broad and diverse circles, often lacking close friends and for them, new relationships often become more important than the old ones.
- **Cognitive and emotional hierarchy**: Stronglinkers are centered on a few ideas and emotions and find it rather difficult to change their cognitive structures. They are highly hierarchical and self-disciplined. Weaklinkers admit many competing ideas and emotions. They have a conflicting and ambiguous hierarchy and tendency toward a spirit of playfulness and instability.
- **Cognitive dimensions**: Stronglinkers can imagine only a few attitudes, whereas their counterparts can imagine numerous attitudes and assess a large variety of interacting people.
- **Internal images of the outside world**: Stronglinkers are generally rigid and usually match new concepts with pre-set notions. Weaklinkers are flexible, highly adaptive to new information and unfamiliar environments.
- **Tolerance for ambiguity**: Stronglinkers – low; weaklinkers – high.
- **Creative behavior**: Stronglinkers – high in structured, hierarchical schemes; weaklinkers – high in diffuse structures.

- **Major problems**: Stronglinkers – loneliness, depression, too rigid, too logical lifestyle; weaklinkers – unfocused life, variable motivations, lack of endurance.
- **Network structure**: Stronglinkers' scale-free networks get closer to a star-like network, whereas weaklinkers' scale-free networks get closer to a random model.

However, conjectures about the aforementioned personality traits still need to be verified in research.

The Process of Social Transformation and the Strength of Ties

The existing literature on strength or weakness of ties relates mainly to the structure of networks at a given time. However, it is also interesting to observe how networks behave over time, especially when transition periods are involved (for example, communities facing natural disasters; families receiving information on a child's disability; or entire rural communities undergoing intense economic development). This chapter puts forward the notion that the strength or weakness of ties varies along with different phases of the transition and that its flows are nonlinear.

Groups or Societies Facing Difficulties

When facing natural disasters those affected experience two tendencies on the part of responders to their plight: mobilization followed by deterioration of received support. In the immediate aftermath of floods, communities usually experience a mobilization of social support among many community members who are not 'first-string' connections. This sort of support, albeit temporary, raises the victims' buoyancy and resilience. However, in the long run this short-lived social capital deteriorates and disaster victims often end up with even fewer social-support networks than they started with; this leaves the former victims deprived of social networks and support, which usually shatters their psychological immune system and increases their tendency to assume the mantle of victim (Kaniasty 2003; Norris et al. 2005).

Similar tendencies may be observed within families of autistic children: in the first phase, after receiving the medical diagnosis the family initially experiences a huge mobilization of support coming from family members, friends, and loosely connected individuals. However, in the long run this social capital dissolves and leaves the families isolated and feeling abandoned (Praszkier 2005); from a longer perspective, a new occurrence of social capital may arise, comprising distant family members and acquaintances (Wroniszewski 2010).

In both cases, in the initial phase there is a rapid escalation of weak ties (e.g., temporary help received from distant acquaintances or relatives) followed by a

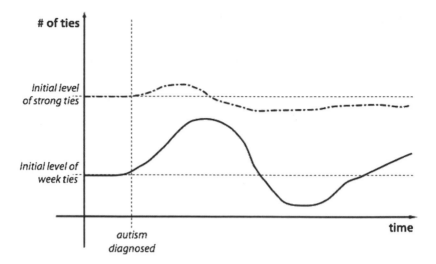

Fig. 7.3 Increase and decrease of weak ties after the autism diagnosis; in the third phase there is a chance for a new social-network buildup; the strong ties remain more or less constant, usually in the long run below the initial level

decrease, even to levels below the initial ones, and, under some conditions, followed by a new phase of a buildup of weak-ties, as illustrated in Fig. 7.3.

Those two examples indicate that in various situations, when families, groups or societies are facing unwanted rapid changes (discontinuities), an unstable situation may occur in a certain period of time, followed by a downward trend or an increase of weak ties.

Societal Transitions

Some groups or societies, however, are predominantly structured around strong ties. For example, some immigrant communities tend to form close, strongly bonded internal structures within their newly adopted culture, which provide support and emotional security through personal confirmation via a shared common identity and sustaining their original cultural heritage through intergenerational transmission. While at first glance, these developments can be positive, in the long run, they also tend to increase isolation and resistance to establishing links with others.

Some underdeveloped communities, similar to the one in Żegocina county depicted in the introduction, remain unchanged, even if the entire region around them is growing and modernizing; they remain deeply attached to antiquated ways and form obsolete and thus marginalized islands of irrelevance. In those 'island' communities weak ties are at a minimal level, often resulting in high unemployment, inadequate educational performance, lack of access to external resources and, as a result, a high level of frustration.

The Żegocina community couldn't transform and succeed without both kinds of links – strong and weak. On the one hand, strong links were the backbone of close-knit circles. On the other, building an enterprising, vibrant community would not be possible without weak ties connecting the close-knit small groups and encompassing the entire community. Csermely (2009) maintains that strong links are important in new democracies, whereas weak links gradually build up the civil society (as in the example of the Żegocina community successfully coping with a natural disaster) and also play a prominent role in maintaining the network's stability.

Families of autistic children, who usually close themselves off (after the initial period of an increase of connections) from the outside world (Glass 2001). They boil down the connections with the outside world to a few weak ties with the helping organizations; internally those families remain strongly bonded (often to a level that can threaten spousal bonds and, by extension, family cohesiveness).

In all these cases a positive transition includes opening the closed communities or families to the outside world, with the increase of new (weak) connections, providing cultural exchange, access to information and resources, new perspectives, and greater access to jobs.[7]

Social Entrepreneurs Add Weak Ties

Social entrepreneurs change the properties of the social system by modifying such parameters as trust and the propensity for cooperation, both of which build social capital (Praszkier et al. 2009). Social capital becomes the bedrock of the new social context, which influences people's mindsets and attitudes to the point that they are more prone to communicating, sharing, cooperating and becoming receptive to adopting creative solutions. The last is especially important, because intractable social problems can be successfully addressed only with an innovative approach.

One component usually required for innovative solutions is the building of new bonds, both inside the target group or society (which augments the development of social capital) as well as outside the group, as a way to provide access to information and resources, as well as to close educational and developmental gaps. The new bonds (with the outside world) are of a weak kind, as is revealed by the lower amount of time, emotional intensity, intimacy and reciprocation involved as compared with the strong ties they maintain within their 'old' clusters. Below are two illustrations of this process.

---

[7] For the latter, see Granovetter (1995).

Krzysztof Margol, Combating Unemployment in Disadvantaged Communities

Krzysztof Margol[8] is contributing to the revitalization of Poland's underdeveloped rural areas by stimulating community awareness of business opportunities and providing new jobs to the rural unemployed. Through the NIDA foundation[9] that he founded, Krzysztof Margol is training the unemployed to start new business ventures. He also convinces businesses, banks, and local governments to back new job-creating initiatives.

He has been doing so by stimulating new connections among inhabitants through a variety of forums, and through creating more than 40 new civic organizations. Krzysztof Margol also launched the online youth radio, which creates several platforms for promoting young people's connections. A variety of external ties were created, connecting the community members with communities in other regions and with various organizations in Poland, Denmark, Ukraine and Russia.

The new local initiatives are attracting tourists, are sparking new relationships and are encouraging community members to visit other regions in order to study other successful projects, engendering yet another portfolio of connections.

Dr. Michał Wroniszewski, Creating a National System for Treating Autism

Dr. Michal Wroniszewski[10] is creating a national movement to accept disabled or 'special-needs' individuals as integrated members of society in Poland, with a primary focus on autistic children and their families.

The Synapsis Foundation[11] he heads has initiated several regional and national initiatives and sparked an array of civic organizations in order to create a movement that addresses all the stages from prenatal and post-natal screening to kindergardens for autistic children, and sheltered workshops for autistic adult persons to obtain work and support themselves. This movement includes several mutual-support and information-exchange forums for families (for example, Mothers' Forum, or Club Forum), for professionals and volunteers.

These two examples illustrate how social entrepreneurs initiate a change process through building social capital.[12] One of the ways they are doing so is by bringing about new weak connections linking individuals within the target group or community with the external world.

---

[8] See: www.ashoka.org/fellows/krzysztof_margol.

[9] See: www.nida.ecms.pl/index/?lang_id=2.

[10] See: www.ashoka.org/fellows/michal_wroniszewski.

[11] See: www.synapsis.waw.pl.

[12] For social capital built by social entrepreneurs see: Praszkier et al. (2009).

Dynamical Model: Analogies to the Dynamics of Hydrogen Bonds

Hydrogen Bonds: Dynamical Account

The following analysis expands upon Granovetter's comparison of the dynamics of social ties to the way certain molecules 'relate' to each other. As previously noted, Granovetter contends that groups or societies with no or very few weak ties are analogous to the gas state of water vapor in which H_2O molecules are all separated, drifting individually until the vapor cools, at which time the molecules associate with each other through weak hydrogen bonds and are no longer free-floating, becoming the liquid we recognize as water. In other words, the act of cooling the vapor coerces a phase transition into the liquid (water) state. Further cooling provokes the next phase transition, into the solid (ice) state, with the previously weak bonds becoming strong and holding together the solid structure (see Fig. 7.4). The strong (covalent) bonds holding the water molecules together remain at more or less the same level[13] during those phase transitions.

Groups, Families and Societies: A Dynamical Model

Harmony Between Weak and Strong Ties

This model fills in the analogy with social dynamics: the most adaptive and 'healthy' state is characterized by the coexistence of strong and weak bonds, both playing a significant role in building the balance between cohesion and openness of groups, communities or societies. Weak links stabilize all complex systems: they provide the 'small-world' atmosphere, enable synchronization of bottom-up networks, provide communication channels, ensure that the relaxation of the stressed networks goes smoothly, and the network is integrated and behaves as a whole (Csermely 2009). Strong links provide the backbone, help in maintaining and transmitting values and traditions, provide a sense of identity and are reference points in case of disturbances. Following through with the analogy, we can characterize this harmony between strong and weak links as the liquid state. This harmony can, however, breakdown, through a transition phase, into a state where the balance of weak and strong ties is lost.

---

[13] It takes a much higher temperature to break apart the molecules into their constituent single atoms, breaking the strong (covalent) bonds. At still much higher temperatures, the atoms are torn apart into nucleus and separated electrons.

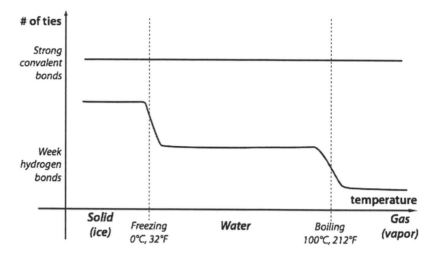

Fig. 7.4 The water phase transitions and the strength of bonds

Atrophy of Weak Ties

At one end, the transition phase may symbolize a political or social upheaval, such as a riot, a conflict, a revolution, a loss of a major source of income (e.g., when the only industry providing most of the jobs goes bankrupt), or, in the case of a family – an acute conflict. The upheaval or destructive and protracted conflict situation may result in the atrophy of weak ties at which point conflicted families turn into isolated fortresses, small, close-knit cliques in conflicted communities (see Fig. 7.5).

Too Many Weak Ties

At the other end, too many weak ties create an overload of connections, so that gradually relationships threaten to fall apart due to lack of attention (Borgatti et al. 1998). The predominance of weak ties over strong ones creates the prospect of losing the basic support circles and the links becoming detached from their roots. Moreover, there is the possibility that the excess of weak ties may at some point supplant the few but necessary strong ties. As a result, the effective social capital withers as the multiple weak ties leave no space for cultivating the strong ones (see Fig. 7.5: in the left section, strong bonds decrease as a reaction to the increase of the weak ones). In the case of families, cultivating too many weak ties may disperse time and energy over stronger social networks, lessening the capacity for growing internal family bonds (the **bridging** kind of social capital becomes predominant over the **bonding** type (Putnam 2000)). In this state, the society may turn into an amorphous soup.

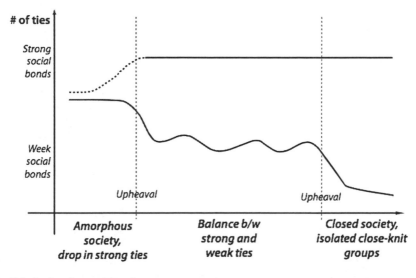

Fig. 7.5 Society in transition phases

Endnotes

It has been demonstrated that one of the components of the endogenous change process facilitated by social entrepreneurs regulates the important control parameter – the strength or weakness of ties. The goal is to strike a balance between strong and weak ties, which in some situations means 'opening windows' to the external world, and in other situations translates to restoring the cultural and historical identity of a particular group. In the first case, social entrepreneurs foster a process leading to an increase of weak ties; in the latter, they add the strong ones.

Using the water analogy, social entrepreneurs often either cool down (i.e., increase the number of weak ties) or warm up (i.e., decrease the ratio of weak-to-strong bonds) the target populations in order to achieve the optimal balance (the 'liquid' state). They manage to increase weak ties in, for example, isolated and backward communities, or in isolated families in post-traumatic distress. On the other hand, they develop ways to create strong ties, working with youth groups whose members are detached from their families and cultural roots and have multiple weak ties and are 'floating' across the society, with no feeling of belonging (such groups are easily attracted to and abused by extremists, as there is a psychological need for identification, and they are driven by the notion that any identification leading to regaining the feeling of belonging is better than none).

Restoring a healthy balance between weak and strong ties may be one of the methods of bringing durable change to disadvantaged groups or communities. It would be worth further study, which may show that focusing on the increase of weak ties (which is relatively cost-effective) may be an efficient method for engendering change. Some clues have already appeared in research: for example

Coulson (2010) found that the use of weak bonds through social networks increases the likelihood of engaging in citizen communication or civic participation; also, Sakurai (2008) asserts that weaker ties are more advantageous than other methods in accessing social capital.

It would also be worth studying how the increase of weak ties influences intractable conflict situations: the conjecture is that weak ties may ease tensions and open the conflicted groups in which predominant strong internal bonds keep them isolated and in a permanent state of inter-group conflict (see Bar-Tal 2007).

Finally, the dynamic model of the behavior of water molecules may also serve as a foundation for the computer modeling the process of social transformation, where the number of weak bonds may serve as one of the control parameters.

References

Anonymous Business Editors. Hill and Knowlton and Ashoka form precedent-setting global partnership. Business Wire (2002), htp://findarticles.com/p/articles/mi_m0EIN/is_2002_June_27/ai_87867887. Retrieved 11 Nov 2010

Ashoka, Innovators for the Public: Selecting Leading Social Entrepreneurs. Ashoka, Arlington (2000)

Barabási, A.L.: Linked. A Plume Book, Cambridge (2003)

Bar-Tal, D.: Sociopsychological foundations of intractable conflicts. Am. Behav. Sci. **50**(11), 1430–1453 (2007)

Borgatti, S.P., Jones, C., Everett, M.G.: Network measures of social capital. Connections **21**(2), 27–36 (1998)

Bornstein, D.: How to Change the World: Social Entrepreneurs and the Power of New Ideas. Oxford University Press, New York (2004)

Bornstein, D., Davis, S.: Social Entrepreneurship: What Everyone Needs to Know? Oxford University Press, New York (2010)

Brinckerhoff, P.C.: Social Entrepreneurship: The Art of Mission-Based Venture Development. Wiley, New York (2000)

Coulson, J.: The Strength of Weak Ties in Online Social Networks: How Do Users of Online Social Networks Create and Utilize Weak Ties to Amass Social Capital? LAP Lambert Academic Publishing, Saarbrücken (2010)

Csermely, P.: Creative elements: network-based predictions of active centres in protein and cellular and social networks. Trends Biochem. Sci. **33**(12), 569–576 (2008)

Csermely, P.: Weak Links: The Universal Key to the Stability of Networks and Complex Systems. Springer, Berlin (2009)

Dees, J.G.: The Meaning of Social Entrepreneurship. Kauffman Center for Entrepreneurial Leadership, Kansas City (1998), http://www.caseatduke.org/documents/dees_sedef.pdf. Retrieved 21 Sep 2012

Drayton, W.: The citizen sector: becoming as entrepreneurial and competitive as business. Calif. Manage. Rev. **44**(3), 120–131 (2002)

Drayton, W.: Where the real power lies. Alliance **10**, 29–30 (2005)

Elkington, J., Hartigan, P.: The Power of Unreasonable People. Harvard Business Press, Boston (2008)

Gendron, G.: Flashes of genius: interview with Peter Drucker. Inc. Mag. **18**(7), 30–39 (1966)

Gentile, M.C.: Social Impact Management and Social Enterprise: Two Sides of the Same Coin or Totally Different Currency? Aspen Institute for Social Innovation in Business, New York

(2002), www.aspeninstitute.org/sites/default/files/content/docs/business%20and%20society%20program/SOCIMPACTSOCENT.PDF. Retrieved 11 Nov 2010

Gladwell, M.: The Tipping Point: How Little Things Can Make a Big Difference. Back Bay Books, Boston (2002)

Glass, P.W.: Autism and the family: a qualitative perspective. Doctoral dissertation submitted to the Faculty of the Virginia Polytechnic Institute and State University (2001)

Granovetter, M.S.: The strength of weak ties. Am. J. Sociol. **78**(6), 1360–1380 (1973)

Granovetter, M.S.: The strength of weak ties: a network theory revisited. Sociol. Theory **1**(1), 201–233 (1983)

Granovetter, M.S.: Getting a Job: A Study of Contacts and Careers. University of Chicago Press, Chicago (1995)

Hammonds, K.H.: A lever long enough to move the world. Fast Company, **90**, 60–63 (2005) Available on: http://www.fastcompany.com/52233/lever-long-enough-move-world. Retrieved 21 Sep 21 2012

Kaniasty, K.: Klęska żywiołowa czy katastrofa społeczna? Gdańskie Wydwnictwo Psychologiczne, Gdańsk (2003)

Krackhardt, D.: The strength of strong ties: the importance of Philos in organizations. In: Nohria, N., Eccles, R.G. (eds.) Networks and Organizations: Structure, Form and Action, pp. 216–239. Harvard Business School Press, Boston (1992)

Leadbeater, C.: The Rise of the Social Entrepreneur. Demos, London (1997)

Lin, N.: Social Capital: A Theory of Social Structure and Action. Cambridge University Press, Cambridge (2001)

Mair, J., Robinson, J., Hockerts, K.: Introduction. In: Mair, J., Robinson, J., Hockerts, K. (eds.) Social Entrepreneurship, pp. 1–13. Palgrave MacMillan, New York (2006)

Noble, T.: Social Theory and Social Change. Palgrave, New York (2000)

Norris, F.H., Baker, C.K., Murphy, A.D., Kaniasty, K.: Social support mobilization and deterioration after Mexico's 1999 flood: effects of context, gender, and time. Am. J. Community Psychol. **36**(1–2), 15–28 (2005)

Nowak, A.: Dynamical minimalism: why less is more in psychology? Pers. Soc. Psychol. Rev. **8**(2), 183–193 (2004)

Petróczi, A., Nepusz, T., Bazsó, F.: Measuring tie-strength in virtual social networks. Connections **27**(2), 39–52 (2007)

Porter, J.: Weak ties and diversity in social networks. Bokardo Social Web Design (2007). Available on: http://bokardo.com/archives/weak-ties-and-diversity-in-social-networks/, 5 Oct 2007

Praszkier, R., Nowak, A.: Social Entrepreneurship: Theory and Practice. Cambridge University Press, New York (2011)

Praszkier, R.: Family social capital and autism. A lecture for the Synapsis Foundation. Synapsis Foundation, Warszawa (2005, unpublished material)

Praszkier, R., Nowak, A., Zablocka-Bursa, A.: Social capital built by social entrepreneurs and the specific personality traits that facilitate the process. Psychol. Spoleczna **4**(10–12), 42–54 (2009)

Putnam, R.D.: Bowling Alone. The Collapse and Revival of American Community. Simon and Shuster, New York (2000)

Steyaert, C., Hjorth, D.: Introduction: what is social entrepreneurship? In: Steyaert, C., Hjorth, D. (eds.) Entrepreneurship as Social Change, pp. 1–18. Edward Elgar Publishing, Cheltenham (2006)

Wroniszewski, M.: Some conclusions from observing the autistic families in longer perspective. Synapsis Foundation, Warszawa (2010, unpublished material)

Chapter 8
A Dynamic Approach to Multicultural Integration

Wouter E. de Raad

Abstract In this chapter a dynamic model of multicultural integration is presented, formalized and tested in a series of computer simulations. Three main points are made concerning: First, how simulations aid in the analysis of the logical consistency and premises of a theoretical model, in this case Berry's acculturation model, on which our dynamic model is built. Second, how computer simulations may help in development of new theory by producing outcomes which can serve as testable claims. Third, the social mechanisms, that lie at the basis of a limited number of scenarios of multi-cultural integration.

The multicultural make-up of societies, especially the coexistence of people with different ethno-cultural backgrounds, is one of the major concerns of our times. Recently we have been confronted with many examples showing that cultural differences can and do lie at the basis of conflict. Violence and incidents on a smaller or larger scale have been observed in many countries, on different continents. As a rule societies struggle with issues related to multiculturalism, seemingly without exception. In several countries the popularity of political parties that base their views on multicultural intolerance seems to be growing (Rydgren 2007). Clearly people are concerned with the multicultural make up of their societies; something that is only likely to become more pronounced as migration is increasing and so the number of countries affected is growing (Castles and Miller 2009). But multiculturalism can also be an issue of concern in situations where migration does not play a dominant role. Examples of outbursts of violence in Iraq, India, Pakistan, Sri Lanka, Nepal, the Philippines, Sudan, Somalia, and Kyrgyzstan (to name some examples that have attracted considerable international attention) make us aware that multiculturalism can also be an issue in instances not directly related to migration. Differences of values and goals may be rooted in, for instance,

W.E. de Raad
Faculty of Psychology, University of Warsaw, ul. Stawki 5/7, 00-183 Warsaw, Poland
e-mail: wderaad@psych.uw.edu.pl

A. Nowak et al. (eds.), *Complex Human Dynamics*, Understanding Complex Systems, 131
DOI 10.1007/978-3-642-31436-0_8, © Springer-Verlag Berlin Heidelberg 2013

differences in religion, political ideology, ethnicity, or history; but at a more general level it seems that those conflicts are related, at least in part, to differences in culture. Clearly, the 'objective' cultural differences are of less importance than the cultural differences as perceived by the parties involved. In contexts such as these, it may be helpful to use a rather broad understanding of what constitutes cultural differences and also look at differences between subcultures within groups that could be considered to be of the same culture and ethnicity. The large scale conflict between Hutu's and Tutsi's in Rwanda in the mid nineties may well be an example to which this would apply (Bowen 1996; Fearon and Laitin 2000).

What characterizes all multicultural societies is that they involve social processes of people with cultural differences, whether large or small, recent or historic, living together and interacting. How the patterns of interaction between people will evolve, and if (and if so, how) those patterns can be guided are important questions. Research on the relationship between social capital and conflict shows the importance of social contacts, patterns of interaction, and changes in those patterns (Colletta and Cullen 2000; Putnam 1993). If we assume that an objective for societies is peaceful coexistence, then social integration of different groups – the establishment and maintenance of social ties between people, a central aspect of social capital, is a key issue. Through social ties people receive information about topics as diverse as job opportunities, language, customs, norms, where to obtain certain items, where to obtain information, etc., which are clearly important to people new in a society. Also, contact helps to reduce prejudice and fear and increases empathy (Pettigrew and Tropp 2006, 2008) both of which are important in bringing people closer together socially. Knowing that the establishment of social ties through interaction is central for multicultural societies, how can we capture this theoretically, both to gain a deeper understanding and to be able to make reliable predictions?

Research on ethno-cultural contact has shown that contact may foster positive outcomes (Pettigrew and Tropp 2006, 2008), but a question remains: what makes people from different cultures seek contact in the first place? Berry's model of acculturation tries to answer that question. According to Berry (1980, 1997), people with a different ethno-cultural background than the majority – designated as the non-dominant group – are faced with two important questions: to what to extent to maintain their heritage culture and identity, and to what extent to seek contact with and participate in their 'host' society? Based on the combination of preferences people can have on those two issues, Berry defined four distinct acculturation strategies:

- An assimilation strategy, where individuals do not wish to maintain their cultural identity, but prefer to maintain contacts with another culture.
- A separation strategy, where individuals prefer to hold on to their original culture and want to avoid contact with the other group.
- Integration, when a person is interested both in maintaining their original culture and in seeking interaction with the other culture.
- Marginalization, when, in contrast, there is neither interest in cultural maintenance nor in cross-cultural interaction.

Berry duplicated the framework of the non-dominant group acculturation strategies for the dominant group: when the dominant group would pursue assimilation of the non-dominant group this would be called the 'melting-pot' strategy; if integration would be preferred this is termed 'multiculturalism'; separation from the side of the dominant group is termed 'segregation' and marginalization is labeled 'exclusion.'

Berry points out that the resulting situation of groups being in contact depends not only on their acculturation strategies, but also on the combination of the strategies held. For instance, if the non-dominant group prefers integration and the dominant group prefers exclusion it is clear that a well integrated multicultural society will not easily result. The issue of compatibility of acculturation strategies led Bourhis and colleagues (1997) to propose an interactive model of acculturation in which they take the specific combinations of acculturation strategies into account, and make predictions about the societal outcome of those combinations, which they label 'consensual', 'problematic' or 'conflictual.'

In these models the way people behave in a multicultural society is a reflection of their acculturation strategies and the resultant situation in a society is influenced by their combination. Although Bourhis et al. (1997) make predictions about which outcomes are related to specific combinations of acculturation strategies, it is unclear exactly why and how those outcomes are related to the various combinations of strategies. So, although these models identify and describe important variables and phenomena of multicultural societies, to understand why a multicultural society remains segregated or becomes integrated, a deeper understanding of the social processes that take place is needed.

Both the above models acknowledge the personal dimension of acculturation, but when talking about outcomes focus mainly on groups. For descriptive purposes, and for the sake of clarity, it may make sense to talk about groups and inter-group relations in a multicultural society, but it should be clear that it is individual people that adapt and integrate, not groups. Individuals engage in social interactions and make decisions that, at an aggregate level, can be interpreted as societal outcomes. So, to gain a better understanding of the processes underlying multicultural integration we need to switch attention from the group to the individual level.

Understanding aggregate level outcomes based on individual level behaviors is not an easy task. The social systems we try to understand are complex and dynamically evolving processes consisting of numerous interacting elements. Complex systems such as these are notoriously difficult to analyze and understand (Waldrop 1992). Social psychological theories generally specify fixed-in-time causal relationships and seem unlikely to offer much insight in dynamic social processes. In this chapter we will propose a dynamic model of multicultural integration that will provide the insight that links individual level behaviors and social interactions to group level of societal outcomes.

Dynamic systems are characterized by ongoing interactions between elements; outcomes at an aggregate level are often non-intuitively related to variables at the level of the individual elements. To analyze such systems we need to make use of tools appropriate for dealing with this complexity. Computer simulation has emerged as a method par excellence for doing this. Initially stemming from the natural sciences, computer simulation is also a valuable tool for social psychology (see this volume, Nowak et al., Chap. 1).

One obstacle to the development of dynamic models is that current social theories often lack the precision required to formulate the rules or algorithms so that they can be used in a computer simulation. However, existing theories and models can often be adapted, or can serve as inspiration for simulation studies. The key is to choose the necessary variables to explain the phenomenon of interest, while keeping the number of variables included to a minimum, a principle Nowak (2004) termed dynamical minimalism.

A well known example of such a model is Schelling's dynamic model of segregation (Schelling 1971). With this model he showed that simple individual rules with regard to preference of racial make-up of one's direct environment can lead to surprising group level outcomes. For instance, if individuals would move from their current location to a random empty location when they are surrounded by more individuals of a different type to themselves, then complete segregation of the groups would result. Schelling developed his model to explain interracial segregation, but along similar lines the results can be adopted to explain segregation of people from different cultures. Although this model explains why individuals from different cultures may live in separation, it seems unlikely that people's behavior would depend on the actual rules specified in this model.

A different approach, adopted by Mimkes (1995), uses binary alloys as a model for a multicultural society. His model is based on the laws of thermodynamics and is truly dynamic; the final state of the system depends on the interactions of the agents (or particles). The behaviors of people in this model are based on the rules or natural laws that govern the behavior of physical particles. Mimkes outlines striking similarities between alloys of metals at the molecular level (solubility of molecules) and multicultural societies (patterns of intermarriage between people from different races or religions). Despite the parallels, the precision of the model and its elegant simplicity, there is a fundamental problem with the rules that govern the behavior of the elements in this model. We know that people behave in a different fashion than molecules, and close inspection of the principles defining the behavior of the elements in the model shows that those principles are not in line with findings from social psychology: The way people are attracted to each other differs from the way molecules are attracted to each other. So even though the outcomes of this model may be comparable to societal situations, the rules and dynamics that produce these outcomes do not compare to social dynamics and are unlikely to provide the understanding that we seek. What we need is a model that combines the precision and dynamic nature of Mimkes' model, but is firmly based on an empirically supported social psychological theory, as in Berry's model.

In this chapter we develop a theoretical model that is based on Berry's model of acculturation but is dynamic in nature. We will formalize this model in precise rules so that it can be implemented in a computer simulation. We will analyze and discuss the outcomes of the simulations in the light of multicultural integration. More specifically, we will test a number of claims based on Berry's model and we will look at several distinct scenarios that can be identified in the simulations, discussing their relevance for real life situations. We will also show how the outcomes of these simulations help to formulate new claims which can not only be tested empirically but also help in developing new theories. These claims will be discussed in the light of the existing knowledge of multicultural integration. We will conclude the chapter by discussing the shortcomings of this model and by suggesting steps that could be taken to further develop it.

Developing the Model

The central premise of Berry's model is that people posses acculturation strategies that are specifically related to how to live together in a multicultural setting. As such, Berry's model not only perfectly fits the topic of our interest but also, by linking attitudes to behavior, opens up the possibility for defining precise rules which are needed for specifying a formalized model. For our dynamic model we need to specify precisely how people's strategies or attitudes are related to their behaviors, in order to know how these will affect interpersonal interaction and the establishment of social ties.

In his model Berry makes a distinction between dominant majority groups and non-dominant minority groups. For the sake of simplicity we will focus on situations where there are only two groups.

Acculturation strategies of minority or non-dominant groups are defined by their attitudes towards both culture maintenance and contact with the majority group and culture. For our purposes, we interpret those two dimensions in the following way. The preferences for cultural maintenance and retention of the ethno-cultural identity will be seen as a relative preference to interact with people from their own ethno-cultural group and therefore is a relative preference for seeking contact with individuals of their own ethno-cultural group. Conversely, a preference not to maintain culture is seen as a preference not to have contact with individuals from their own group. For the contact with the majority group dimension a similar logic is followed. A preference to have contact with and to participation in the majority, dominant culture is interpreted as a relative preference for contact with majority members, and vice versa.

Berry also defined the acculturation strategies of the majority group in relation to the acculturation of the minority group. For instance, if a person from the majority group prefers integration of minorities, this acculturation strategy is termed 'multiculturalism.' Although majority members have a preference for a specified way of acculturation, from the definition it seems this preference mainly concerns how minority members should behave, not their own behavior. If a majority member prefers multiculturalism, i.e. they prefer minorities to integrate, the related question should be phrased as, for example, "Do you prefer minority members to seek contact with and participate in your (majority) culture?" This is a passive formulation and quite different from the corresponding question put to minority members: "Do you want to seek contact with and participate in the majority culture?"

Majority attitudes and behaviors seem to be distributed over a more limited range than is the case with the minority group. A majority's preference for the minority group to seek contact (versus a preference for them not to seek contact) would seem to result in less pronounced, rather passive behavioral consequences, probably comparable to about the middle range of the same scale of the minority, who have a more active approach ranging from active contact seeking to active contact avoidance.

No reason is offered as to why the contact dimensions should be different for majority and minority members, and it seems an unnecessary complication of the model. For our purposes we want to treat both groups identically and so for the majority groups as well, we assume attitudes towards contact with people from a different ethno-cultural group.

A second difference between the majority and minority group regards the culture maintenance dimension. Minority groups evaluate their own culture and have to decide whether or not to maintain it. The acculturation strategies of the majority do not include an evaluation of their own culture, only a preference for how minorities evaluate their culture. Although it has been shown that majority members do have a preference for whether or not minorities maintain their own culture, what matters here is how this would change the way that majority members interact with minorities. Since this is already captured by the contact dimension, it seems redundant to have a second dimension that also impacts majority members' preference for contact. We acknowledge that the issue may be an important one, but the question of how the preference for minority culture maintenance influences the preference for contact is not one that needs to be answered here. To solve the issue of the majority culture maintenance dimension and to keep the model as simple as possible we specify that majority members, just as minority members, also evaluate their culture and identity, which influences their preference for contact with members of their own group in the same way as it does for minority members. To summarize, we end up with a very lean model that has the same principles for all individuals, majority or minority. People have preference about contact with people from the other group and from their own group, which influences them in selecting their social contacts.

The principles specified so far rest upon several implicit assumptions that need to be made explicit. We assume that humans are social animals and have an innate drive for social contact, which is related to their subjective well being (Baumeister and Leary 1995), which we hereafter will refer to as 'happiness'. We further assume that this innate drive for belonging motivates people into action to increase their happiness (Maslow 1954). In a multicultural setting people's attitudes towards the different groups will influence with whom contact is either sought or avoided so as to increase happiness. In line with Berry, we regard these attitudes as continuous dimensions ranging from positive to negative. These are the basic building blocks needed for a computer simulation.

Now that we have identified the crux of the model, we turn to issues still in need of attention. Berry focuses exclusively on acculturation strategies and not on the underlying dimensions separately. Although we specified how each dimension is

related to distinctive behavioral counterparts, this does not mean that the acculturation strategies – specific combinations of values on those dimensions – should not
be at the focus of our attention. Indeed, the question is how the values on the two
dimensions of attitude, towards own group and preference for contact with the other
group members, are related to measurable outcomes, e.g., degree of multicultural
integration, or ratio of in-group and out-group contacts. By focusing on acculturation strategies instead of the underlying dimensions, Berry seems to suggest that
there is something unique about these strategies that is not captured by the individual dimensions.

This observation can be used to formulate several testable claims. First, since it is
the sign of the attitudes that defines the acculturation strategies, the sign of an attitude
should be more important than its strength. In other words it does not matter if the
value of an attitude is 0.25 or 1 as long as it is positive. However, a negative value,
even if small, would be expected to lead to noticeably different outcomes. Second, for
a given acculturation strategy, different strengths of attitudes should lead to comparable results. Consider the following example: the integration strategy is defined by a
positive attitude towards contact with both the own group and the other group. On a
scale from -1 to $+1$, the attitude towards the own group could be $+1$ and the attitude
towards the other group could be $+0.25$ – this fits the definition of integration. In this
case there is a pronounced preference to have contact with members of the own group
over members of the other group. If the values of the attitudes would have been
reversed, and so more positive towards the other group ($+0.25$ to own group, $+1$ to
the other group) one would still expect the same or similar outcomes, because both
combinations of attitudes would be seen as integration. Third and last, the focus on
acculturation strategies seems to imply that neutral attitude values are of no importance in understanding multicultural integration, or offer no additional information to
the information provided by the acculturation strategies.

Formalizing the Model

For the model to be tested using computer simulations all the relations between the
variables have to be precisely specified – precisely enough for a computer to work
with. We posited that people seek social contact with others to increase their
happiness; happiness therefore is a function of the number of social contacts.
How contacts with others are valued depends on a person's attitude towards the
group to which that contact belongs; a positive attitude means that contact with that
person increases happiness and vice-versa. Although happiness increases with the
number of positively valued contacts, this increase is not linear. In accordance with
the law of diminishing marginal utility (Edwards 1954; Samuelson and Nordhaus
2005) the increase in happiness for each additional (identical) contact becomes
smaller. Specifically we chose for the rule specified in the Dynamic Theory of
Social Impact (see this volume, Nowak et al., Chap. 1), which states that the impact
of a group of individuals increases with the square root of the number of individuals

exerting impact. Because our focus lies on interactions with members from different groups, contacts from each group are counted separately. The relationship between those variables can be summarized by the following formula:

$$H = A_{own} * \sqrt{N_{own}} + A_{other} * \sqrt{N_{other}}$$

Explained in words, this formula expresses that happiness H depends on the square root of the number of contacts from the own group, N_{own}, and from the other group, N_{other}, added together, taking into account the attitude one has towards contacts with each of those groups, A_{own} and A_{other} respectively. Now that we have translated the model into a language that a computer can use, there still remains the question how to represent in a virtual way the social reality in which people exist.

What matters most in multicultural integration is the social connections that exist between people. If we would create a visual representation of a group of people with the existing ties between them, we would end up with a network of interconnected elements. It therefore makes sense to use a simulation model in which people are represented in a network. We chose to represent networks on a 2-dimensional space or grid where individuals or agents could interact with a limited number of others at a given time. This type of simulation belongs to a class of models called 'cellular automata,' which is particularly well suited for simulating dynamic social processes with locally interacting agents (Gilbert and Troitzsch 2005; Hegselmann 1998; Schelling 1971). A big advantage of cellular automata is the ease with which they can be visualized and so observed and analyzed.

Social interactions were simulated using a square lattice of cells (a very large checker board) in a two dimensional space. Individuals, or agents, were assigned to a group, and given an attitude towards their own and other group ranging from −1 to +1. Next they were located to a random cell on the board. Cells could only be occupied by one agent and some cells would remain vacant to allow for freedom of movement of the agents. During the simulation agents would be randomly selected and asked to evaluate their current happiness based on their immediate neighbors, according the formula specified above. Immediate neighbors are agents that are located in any of eight surrounding cells (to ensure that all locations had an equal amount of neighboring cells, the board was turned into a torus, which visually resembles a donut shape). Happiness is dependent on the number of contacts, not on the number of cells; if some of the surrounding cells are empty, happiness is based on less than eight contacts.

Next agents would be offered random vacant locations to move to if they so desired. A decision to move would be made if a spot would provide a higher level of happiness than the current location. However a 'cost' of moving was included by setting a rule that agents would only move if they could obtain at least a ten percent increase in happiness. This is to stabilize the model a little by preventing agents moving for very small increases in happiness. The rationale is that a decision to take action – move – would only be undertaken if this led to a substantially higher

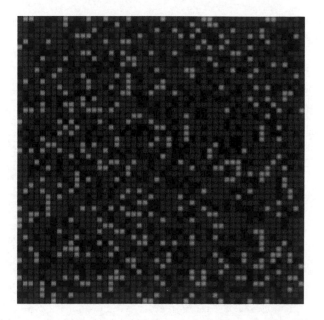

Fig. 8.1 Initial state of a simulation in which agents of two groups – depicted in two shades of *gray* – are randomly distributed on a grid. *Black color* indicates empty spaces to which agents can potentially move

satisfaction. The simulation would end if no agent would move any longer or after an arbitrary set limit.

Simulations

The simulations were conducted on a square grid consisting of 50 rows and columns; 25 % of the 2,500 cells are left open – empty, to provide ample space for agents to move into. There were just two groups, the majority and minority groups, in the ratio 80/20. At the onset of a simulation agents were randomly assigned a location on the grid, which would typically look like Fig. 8.1.

To test the claims regarding acculturation strategies specified earlier we systematically varied agents' attitudes toward their own group and toward the other group. For reasons of simplicity all agents of the same group were assigned the same attitudes; no individual differences were included here. For all members of both the majority and the minority groups, the variables attitudes to own and other group members were each assigned one of the five values: -1. -0.5, 0, 0.5, or 1, systematically varied. This resulted in a $5 \times 5 \times 5 \times 5$ experimental design with 625 conditions. Three simulations were run for each condition, resulting in a total of 1,875 simulations – see here one of the advantages of computer simulation.

The dependent variable was the extent of multicultural integration. This measure was constructed by comparing the actual extent of intercultural contact to the extent of intercultural contact that would be expected on the basis of randomness. In a perfectly integrated multicultural society, culture should not be an issue in interpersonal contacts and thus contacts between the groups should be random. Deviations from randomness can then be used to calculate an index for integration. Our index has a value of zero in case of total segregation – no contacts between the groups – and a value of 1 for perfect integration, or a random distribution of contacts.

We will first focus on the three testable claims that were motivated by the interpretation of Berry's model, before identifying specific scenarios of multicultural integration (or segregation). In our interpretation of the outcomes we will not focus on statistical indicators, but rather on trends that can be inferred from graphs. The reason for this is that data produced by simulations are of a different nature from data typically obtained from social processes in real life. The main difference is that error is nearly absent in simulations unless randomness is explicitly put in the simulations by the researcher. As a consequence results very quickly tend to reach levels of statistical significance, but this does not mean that they are important. Second of all, simulations can be repeated over and over again, so one would always be able to produce statistically significant results if the number of simulations would be large enough. Because we are interested in testing the logical consistency of the model we are more interested in the quality of the outcomes than in the precise quantitative measures.

The first testable claim specified that the sign of the attitudes would be more important than their actual values. In its most extreme form this would mean that all positive and all negative values of an attitude should lead to the same results, but that the differences in outcome between positive and negative attitudes should be rather dramatic. Inspection of the graphs in Figs. 8.2 and 8.3 provide mixed support for this claim. The graphs show how specific values of the attitude dimensions for majority and minority are related to mean levels of integration, averaged over the 375 simulations (three simulations for each of $5^3 = 125$ conditions) related to each indicated value. Figure 8.2a shows the relation between the majority's out-group attitude and the average level integration, for positive and negative attitudes (neutral attitudes are discussed separately under the third testable claim); indeed it shows a sudden jump when values go from negative to positive. In addition the strength of the value also has some influence on the outcome, but not nearly as strong as the sign of the attitude. The attitude towards the in-group has a more linear relation with integration, however, as depicted in Fig. 8.2b, and does not support the claim of the uniqueness of acculturation strategies.

Figure 8.3a, b display the same information as Fig. 8.2a, b, but for the minority group. Although the impact of the minority's attitudes are less pronounced, both attitudes show a 'jump' in levels of integration when going from negative to positive. Here again the precise value of the attitude also has some influence, but these observations do support the notion that the sign of the attitude is more important than its value.

So the notion that the sign of the attitudes is of greater importance than its precise value seems to make sense overall, although the results are not unequivocal.

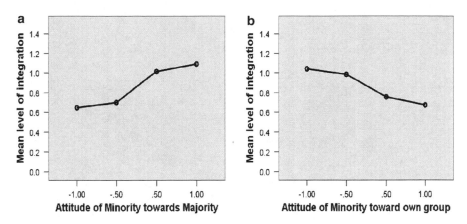

Fig. 8.2 (a) The relation between integration and attitudes of the majority towards the minority; (b) the relation between integration and attitudes of the majority towards their own group

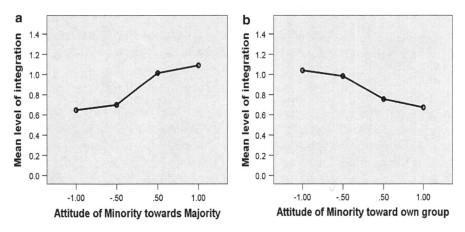

Fig. 8.3 (a) The relation between integration and attitudes of the minority towards the minority; (b) the relation between integration and attitudes of the minority towards their own group

The simulations further suggest that, in addition to the sign, stronger values are related to more pronounced results.

An additional hint that these graphs give us concerns the variable that is of greatest importance in multicultural integration. Looking at the graphs it is clear that the majority's attitude towards the minority causes the greatest variations in integration. It should be realized that majority and minority in these simulations only differ from each other in size; in all other respects they are equal. Hence the difference in impact on integration in these simulations can only be attributed to the numerical superiority of the majority. Intuitively it seems to make sense that the largest group of a population also has the largest influence on the state of that population. This stands in some contrast, however, to tendencies in public debates where frequently the main

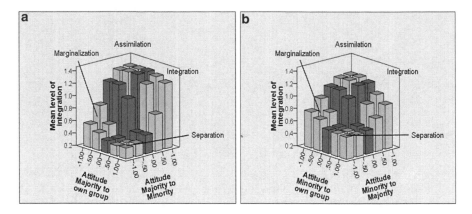

Fig. 8.4 (a) The relationship between combinations of majority attitudes towards the minority and towards their own group and the level of integration; (b) the relationship between combinations of minority attitudes towards the majority and towards their own group and the level of integration

responsibility for multicultural integration is assigned to minorities. This finding, if correct, would have high practical significance and warrants additional future investigation.

The second testable claim that was laid down concerned the different combinations of values of attitudes within a single acculturation strategy, which were predicted to lead to the same levels of integration. Figure 8.4a, b shows the effect of acculturation strategies on the relationship between attitude and integration for the majority and minority groups, respectively. In these figures the acculturation strategies are located in the four corners of the graph, indicated by the lighter columns (the darker columns are not relevant at this point, but will be discussed shortly). Note that because of the difference between this model and Berry's original model we name the acculturation strategies identically for the minority and the majority.

For the separation and assimilation strategies it does seem to be the case that the strength of the attitude has little effect: All the four columns within the strategies are equally high. However, for the integration and marginalization strategies this pattern does not hold very well; the columns differ in height, which means that the specific value combinations within the strategies have a big impact on the level of integration. A reason for this can be sought in combinations of the signs of the values of the attitudes. The prediction of little effect holds if the strategy is made up of attitudes that have an opposite sign – both assimilation and separation combine a positive and a negative attitude. For integration and marginalization, however, the signs of both attitudes are the same: either both positive or both negative. This means that in a way they 'compete' with each other about with whom contact is more sought, or with whom contact is more avoided, and in that situation it does matter which attitude is stronger and which weaker.

So again the support for the notion that acculturation strategies should be at the focus of explanation instead of the underlying attitudes, is equivocal. The picture emerges that, by disregarding information about the continuous values of the attitudes that underlie acculturation strategies, precision is sacrificed with regard to understanding and predicting multicultural integration.

The third testable claim was that neutral attitude values are of no importance. Neutral attitudes do not fit well into acculturation strategies because they exactly fall on the border between them. However, it is plausible for people to have neutral attitudes so the question is what are their consequences. It is difficult to make a prediction how neutral attitudes are related to multicultural integration. In terms of acculturation strategies neutral attitudes are ambiguous because they seem to 'fall in between' two strategies. A positive attitude towards one's own group could, for example in combination with a positive attitude towards the other group, result in the integration strategy, or, in case the attitude toward the other group is negative, result in the separation strategy. However, if the attitude towards the other group is neutral, it is not clear what acculturation strategy we are dealing with, nor which outcomes to predict on the basis of that combination of attitudes. Would the outcomes with a neutral attitude towards the other group fall nicely in between the outcomes of integration and separation strategies, or would they be closer to one than to the other? The dark columns in Fig. 8.4a, b represent outcomes of simulations with neutral attitudes. Note how the darker columns form borders between the acculturation strategies which are located in the corners. As we can observe, the extent of integration related to the neutral values does not neatly fall in between the acculturation strategies that they border, but are closer to one than to the other. The values of the extent of integration seem to cluster around combinations of attitudes with values of opposite sign. In other words, neutral attitudes are more strongly related to assimilation and separation than to integration or marginalization. It can be argued, then, that the neutral values indeed do not offer additional information since their outcomes are essentially captured by some of the acculturation strategies that do not include neutral values. But what is shown by these simulations is that the neutral options are captured by only two acculturation strategies, not four.

After evaluating these three testable claims that were specified in the introduction, we conclude that generally it makes sense to look at acculturation strategies in trying to understand the extent of multicultural integration in a society. At the same time, however, it may make even more sense to look at the specific values of the attitudes that comprise them. If acculturation strategies are based on a combination of two attitudes, then why not look at, and analyze, the influence of their specific values so as to make use of this information that would otherwise be lost? Our simulations have shown that this would lead to a more refined understanding.

Having dealt with these rather theoretical issues, it is worth looking at questions of a more practical nature, such as under which conditions we can expect high rather than low levels of integration and vice-versa. To answer these questions we will look at specific conditions. This simulation study included 625 conditions which clearly is too much to cover in detail, so we will summarize the results by

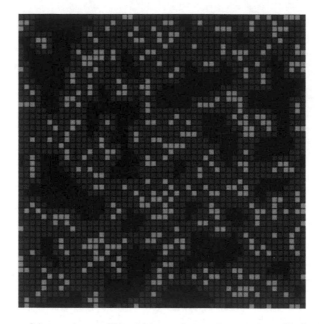

Fig. 8.5 Extent of integration = .993 which nearly equals a random distribution. For both majority and minority out-group attitude = 0.50 and attitude towards own = 1

presenting specific scenarios that capture the outcomes of multiple conditions. We will first limit the scope of our attention a little by leaving out the conditions in which the majority group has negative attitudes towards itself. The reason why we do this is that it simplifies the analysis and that these conditions nevertheless seem to be mostly absent in the general acculturation setting (it seems implausible that a majority group would prefer to let go of its own culture).

What we generally observe is that integration occurs for mutual positive attitudes and segregation for mutually negative attitudes. Simulations producing outcomes of high integration typically look like Fig. 8.5. The resulting pattern is entirely random; group membership does not have an influence on the pattern of contact. This may be compared to people interacting in the United States who are of Italian or Irish descent; it may simply not matter (anymore) to which group one belongs. Another example is the interaction between Catholics and Protestants in the Netherlands. The fact that most people don't know that the southern part of the Netherlands is predominantly catholic and the north mainly protestant illustrates that both groups have integrated so well that this particular make up of the society hardly ever receives any attention.

The opposite is the case illustrated in Fig. 8.6. The groups are completely segregated from each other, with not a single instant of contact between them. Simulations with different negative attitude values show that the resulting patterns are the same regardless of how negative the attitudes are: negative mutual attitudes lead to complete segregation. Although negative attitudes per se do not indicate

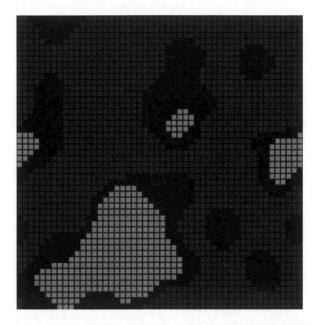

Fig. 8.6 Extent of integration $= 0$ which means total segregation. For both majority and minority out-group attitude $= -0.50$ and in-group attitude $= 1$

societal conflict or violence, it is clear that a situation that is characterized by negative mutual attitudes and lack of contact between groups is at risk of developing into conflict and is potentially violent. Varshney (2001), for instance, shows in detail how the absence of contact and communication is related to violent conflict. He compares how Hindus and Muslims in different cities in India reacted to comparable violence-prone situations. In cases of a relative absence of local interactions between the Hindus and Muslims – both on a day to day basis and on an associative level – violence erupted because there were no mechanisms that could successfully contain or stop the conflict.

The observations that mutual negative attitudes lead to segregation and mutual positive attitudes lead to integration may not be very surprising. In public debates in many countries with a culturally plural population, tolerance is often touted as the key to successful multicultural integration. Tolerance, the acceptance of others who are different, although one may not agree with them, can be represented in our model has having a neutral stance toward the other, or an attitude with value zero. Given the face validity of the claim that tolerance is instrumental in establishing integration, it was much to our surprise that we observed that our tolerant agents did not integrate as well as we expected. As can be seen in Fig. 8.7, tolerance seems to lead to segregation rather than integration.

Segregation in this case is not total, as it would be if mutual attitudes would have been negative, but still the result is striking. It is mainly at the group boundaries that zones of contact between the groups exist. It is of course positive that contact

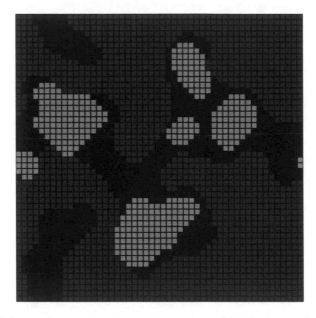

Fig. 8.7 High degree of segregation. Extent of integration = 0.057. In-group attitudes are 1, out-group attitudes are 0

between the groups exists at all, but from these results it would be hard to conclude that tolerance automatically leads to integration. The reason why this pattern emerges is that if people are neutral towards the other group, they will only seek contact with their own. They will not avoid contact with the other, but will not look for it either. In this case segregation is not the result of bad intent, but of indifference. This situation seems to characterize many real life situations; in many places ethnic communities exist that function autonomously from the larger society or from other groups, to a certain extent. We can think of Poles in Chicago, Cubans in Miami, Turks in Berlin, or Algerians in Paris, for example. These results are salient in the light of the earlier discussion about whether the focus should lie on strategies or their underlying attitudes. This example shows that because acculturation strategies do not allow for a neutral position they are limited in their capability of explaining and predicting possible acculturation phenomena.

A situation such as the one sketched in Fig. 8.7, in which two groups inhabit largely different worlds, may balance on a tipping point of failure and success of coexistence. If the contacts that do exist help to establish more contact between the groups, then it is likely that this initial state of tolerance will develop into a more integrated whole. If, however, forces apply that drive the groups apart, and the relatively few contacts that exist diminish, it is not hard to imagine that a situation may occur that is susceptible to conflict or violence. So, although tolerance is obviously to be preferred over disliking, it should be seen as the beginning of the road to successful integration; as a prerequisite, rather than the solution.

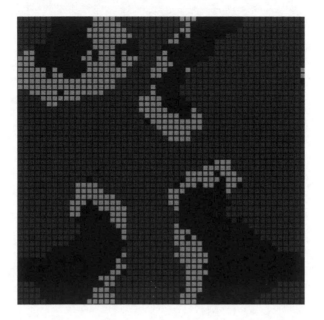

Fig. 8.8 Majority out-group attitude $= -0.25$ minority out-group attitude $= 0.5$. Attitudes towards own group $= 1$

Many contemporary Western multicultural societies are characterized by immigrant groups willing to integrate into the larger society, and thus having at least a moderately positive attitude towards the mainstream group (Berry 1997). On the other hand the current political climate, with the popularity of parties with an anti immigration agenda, suggests that the larger population may often have a slightly negative attitude towards some immigrant groups. Figure 8.8 shows a situation in which the majority group has a mildly negative attitude towards the minority (-0.25) and the minority group has a moderately positive attitude (0.50) toward the majority.

What happens in these situations is that minority members will seek contact with the majority, but because of their slightly negative attitude toward the minority, members of the majority will try to avoid contact with the other and will move away if approached. What results is what may be described as a situation of hide-and-seek, or a chase, in which one group constantly approaches, and the other group constantly withdraws. So, although it seems that integration is established to some extent, this snapshot does not reveal that the simulations has not stabilized into a final configuration and that it never will. In a real social immigration setting this may resemble a situation in which immigrants are motivated to participate in the society through means of work, study or day to day contacts, but are met with general rejection. They will be accepted within their own community, but in the society at large they will not be able to successfully participate, because, as long as the other group actively avoids contact, segregation will result.

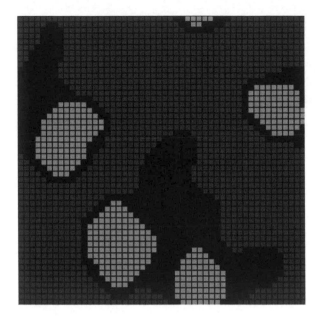

Fig. 8.9 Majority attitude to minority $= -1$, minority attitude to majority $= 0.25$. Attitudes towards own group are 1

These results are relevant in the light of many public debates relating to multicultural integration and the question that is sometimes coined: Who is responsible for integration? For members of the dominant group – the mainstream society – it is easy to say that the responsibility for multicultural integration lies with the immigrants. Our earlier observation when analyzing the relationship between the impact of the attitudes of majority and minority, is here confirmed: the majority group may well play a more decisive role in multicultural integration than the minority group. What Fig. 8.8 teaches us is that even if immigrants are really willing to integrate, a relatively small negative attitude in the society at large may be preventing it from happening. Ironically, the more negative people are towards immigrants, the more emphasis they seem to place on the immigrants' responsibility to adapt or integrate. Figure 8.8 is very clear, however, in showing that responsibility for preventing integration in this case lies with only one group.

In the situation where the majority attitude towards the minority is very negative, -1, and the minority has a slight positive attitude towards the majority (attitudes towards their own group being $+1$) we surprisingly get a state of total segregation, as Fig. 8.9 shows. The majority has such a strong aversion to the minority that they make sure to avoid any contact with them. The attraction of the minority to the majority is not strong enough to make the minority seek contact with the majority at the expense of having to give up contact with some of their in-group members. Inspection of the picture shows that for a minority member to move to the edge, where contact with the other group would be possible, would automatically mean to reduce the number of in-group contacts. In this case a parallel can be drawn with the way in which people often have to negotiate the extent to which they identify with their own people and with the majority group. In a case like this, it seems that the question is either/or, and any sort of dual group membership is obstructed by the negative stance of the majority.

Concluding Discussion

The results of our simulations identify and suggest the existence of a limited number of scenarios of multicultural integration and provide insight in the development of these scenarios over time. If the situation is characterized by mutual negative attitudes, any existing cross-cutting social tissue is likely to disappear, leaving two segregated antagonistic groups. Without contact there will be no communication, which will make the development of social ties virtually impossible. It is problematic that, for any level of mutual negative attitudes, segregation seems to follow. Of course it makes a large difference if groups are only slightly negative or very negative towards each other, but once a movement of segregation has set in, internal group dynamics may propel the situation towards escalation.

In situations where mutual sentiments are positive, groups may integrate. Intergroup contact theory suggests that there are mechanisms that are auto-catalytic in improving attitudes through contact (Pettigrew 1998), and even indirectly through other group members (Wright et al. 1997). The conditions under which these effects are found are unfortunately limited, and reality abounds in examples where contacts are antagonizing rather than bringing people closer together.

Despite of the popularity of the concept of tolerance, simulations with tolerant agents resulted in patterns of surprisingly strong segregation. When people are tolerant and indifferent towards contact with the other group and only actively seek in-group contacts, then segregation will be an unintended side effect. Preaching and fostering tolerance may well make sense in situations of antagonism, where reduction or prevention of conflict is the goal, however by itself it is unlikely to lead to any significant extent of integration. In many societies where frictions between groups exist but there is no real antagonism, the best way forward is the promotion of social ties between groups.

Related to the observations above, even the best intentions from one group are not sufficient to force integration with the other. In a combination of groups with positive attitudes and with slightly negative attitudes, some extent of contact is to be expected. If the attitudes of one group are very negative, however, complete segregation may result even though the other group is mildly positive. These simulations suggest that sustainable integration is only attainable with strong positive mutual attitudes between groups. More information about the dynamics of conflict escalation and resolution and about cognitive aspects of these dynamics can be found in Chaps. 7 and 14 respectively.

All of the above results can and should be empirically verified to further advance our understanding of multicultural integration. The scenarios sketched above – the relation of peoples attitudes (or strategies) to the listed outcomes could all be checked for instance by survey data. Although the results so far are encouraging and already prove to some extent the usefulness of dynamic models and computer simulation in this area, it would be a major step forward if the claims of computer simulation would be empirically verified and used to develop new theory.

We do not want to suggest that this simple dynamic model fully captures the complex reality or even approaches it. The model presented here is clearly an oversimplification of the reality as it presents groups (even literally) in a black and white manner. The fact that because of its simplicity the model stands quite far from reality is both a compliment and a critique. It is a critique in the sense that part of its outcomes clearly do not apply to reality. But it is a compliment in that such a simple model already provides some surprising insights in a complex social phenomenon such as multicultural integration. This model seems to well fit the notion of Nowak's (2004) dynamical minimalism because with very simple rules the model is able to display complex behaviors.

There are several ways in which the model could be improved. First of all the model should be extended and include individual differences in attitudes. Clearly people do not all have the same opinions and representing this in the model would be big step forward. Not only would it be a significant improvement in its own terms, but also it would go beyond the model of Berry and Bourhis and colleagues, who seem to describe multicultural integration mainly at group level. Another thing that should be improved is to make the agents sensitive to the interaction they engage in. The endless hide-and-seek behavior that occurs in some simulations does not occur in reality. People learn from positive and negative experiences and adjust their attitudes accordingly. The inclusion of a mechanism of attitude change could potentially enrich the model enormously. Future modeling attempts have to reveal what this promise holds.

To sum up, this simulation study has shown us three things. Firstly it demonstrates that on the basis of a simple theoretical model with a small set of assumptions, computer simulations allow us to explore and identify various scenarios that provide insight in real life situations of multicultural integration. Because the model is dynamic it makes it possible to uncover some of the social mechanism underlying the phenomena that play an important role in acculturation. Secondly, this study has shown that dynamic modeling makes it possible to test the internal consistency or logic of theoretical models. Berry's exclusive focus on acculturation strategies seems unjustified in the light of the outcomes of these simulations. Third and last, this 'virtual investigation' demonstrated that computer simulations can be used to generate testable hypotheses that can lead to the formulation of new theory. A set of assumptions based on an existing model, led, by simulation, to new hypotheses that could not directly be inferred from the original model. The theories that would be verified (or not) by these predictions are qualitatively different from the initial model that lay at the basis of the simulations. The main accomplishment here is that the new theories take into account the dynamic social processes of interest that are so hard to fathom, because they were 'produced' by those very processes.

References

Baumeister, R.F., Leary, M.R.: The need to belong: desire for interpersonal attachments as a fundamental human motivation. Psychol. Bull. **117**, 497–529 (1995)

Berry, J.W.: Acculturation as varieties of adaptation. In: Padilla, A. (ed.) Acculturation: Theory, Models and Some New Findings, pp. 9–25. Westview, Boulder (1980)

Berry, J.W.: Immigration, acculturation and adaptation. Appl. Psychol. Int. Rev. **46**, 5–68 (1997)

Bourhis, R.Y., Moïse, L.C., Perreault, S., Senécal, S.: Towards an interactive acculturation model: a social psychological approach. Int. J. Psychol. **32**, 369–386 (1997)

Bowen, J.R.: The myth of global ethnic conflict. J. Democracy **7**, 3–14 (1996)

Castles, S., Miller, M.: The Age of Migration: International Population Movements in the Modern World, 4th edn. Palgrave Macmillan, Hampshire (2009)

Colletta, N.J., Cullen, M.L.: Violent Conflict and the Transformation of Social Capital. The World Bank, Washington, DC (2000)

Edwards, W.: The theory of decision making. Psychol. Bull. **51**, 380–417 (1954)

Fearon, J.D., Laitin, D.D.: Violence and the social construction of ethnic identity. Int. Organ. **54**, 845–877 (2000)

Gilbert, N., Troitzsch, K.G.: Simulation for the Social Scientist, 2nd edn. Open University Press, Maidenhead (2005)

Hegselmann, R.: Modeling social dynamics by cellular automata. In: Liebrand, W.G., Nowak, A., Hegselmann, R. (eds.) Computer Modeling of Social Processes, pp. 37–64. Sage Publications, London (1998)

Maslow, A.H.: Motivation and Personality. Harper, New York (1954)

Mimkes, J.: Binary alloys as a model for the multicultural society. J. Thermal Anal. **43**, 524–537 (1995)

Nowak, A.: Dynamical minimalism: why less is more in psychology. Pers. Soc. Psychol. Rev. **8**, 183–192 (2004)

Pettigrew, T.F.: Intergroup contact theory. Annu. Rev. Psychol. **49**, 65–85 (1998)

Pettigrew, T.F., Tropp, L.R.: A meta-analytic test of intergroup contact theory. J. Pers. Soc. Psychol. **90**, 751–783 (2006)

Pettigrew, T.F., Tropp, L.R.: How does intergroup contact reduce prejudice? Meta-analytic tests of three mediators. Eur. J. Soc. Psychol. **38**, 922–934 (2008)

Putnam, R.D.: The prosperous community: social capital and social life. Am. Prospect **13**, 35–42 (1993)

Rydgren, J.: The sociology of the radical right. Annu. Rev. Sociol. **33**, 241–262 (2007)

Samuelson, P.A., Nordhaus, W.D.: Economics, 18th edn. McGraw Hill, Boston (2005)

Schelling, T.C.: Dynamic models of segregation. J. Math. Sociol. **1**, 143–186 (1971)

Varshney, A.: Ethnic conflict and civil society. World Polit. **53**, 362–398 (2001)

Waldrop, M.M.: Complexity: The Emerging Science at the Edge of Order and Chaos. Simon & Schuster, New York (1992)

Wright, S.C., Aron, A., McLaughlin-Volpe, T., Ropp, S.A.: The extended contact effect: knowledge of cross-group friendships and prejudice. J. Pers. Soc. Psychol. **73**, 73–90 (1997)

Chapter 9
Social Change Initiated by Social Entrepreneurs

Agata Zabłocka-Bursa and Ryszard Praszkier

Abstract In this chapter we introduce various aspects of social change. The departure point is an overview of social change theories, highlighting Kurt Lewin's 'force fields' change theory and Andrzej Nowak's 'bubbles' theory of social change. Next, we describe the phenomenon of social entrepreneurship and the role of the individual in introducing social change. Then we explore what are the specific methods used by social entrepreneurs that, by enhancing trust and the propensity for cooperation, empower their communities. Also we present field research analyzing the results achieved by social entrepreneurs, especially their impact on the people and societies in the places where they operate. We conclude that social entrepreneurs launch change process that are multi-pronged and that have long-term results on the individual as well as societal levels.

Introduction

Social change is a fascinating topic attracting increasing interest. Mystery is involved and there is unknown ground to be explored: how does it happen that very huge investments aimed at changing society often yield nearly no results? And, vice versa, why sometimes do small initial impulses trigger a change process resulting in an immense impact? Issues around social change feature in many recent accounts. However, while social change theories abound, research into this process remains uncommon (McLeaod and Thomson 2009). This chapter is a contribution to our understanding of the mechanisms through which social change takes place;

A. Zabłocka-Bursa
Warsaw School of Social Science and Humanities, Chodakowska Street 19/31, 03815 Warsaw, Poland
e-mail: agata_zablocka@wp.pl

R. Praszkier
University of Warsaw, Institute for Social Studies, Stawki 5/7, 00-183 Warszawa, Poland
e-mail: ryszardpr@gmail.com

A. Nowak et al. (eds.), *Complex Human Dynamics*, Understanding Complex Systems, 153
DOI 10.1007/978-3-642-31436-0_9, © Springer-Verlag Berlin Heidelberg 2013

it focuses on a special kind of change process, that which is set off by social entrepreneurs.

Social entrepreneurs target underserved, neglected or highly disadvantaged populations and aim at transformations that yield large-scale benefits (Martin and Osberg 2007). Usually they address seemingly unsolvable social problems generating a huge impact with minimal investments; this they achieve by introducing innovative solutions, initially having no resource other than their social passion, creativity and entrepreneurial frame of mind. Metaphorically, this may appear to be creating 'something out of nothing.' They often do this by triggering a bottom-up process involving and empowering society as a whole. Peter Drucker once said that social entrepreneurs change the performance capacity of society (Gendron 1966), meaning that the impact of social entrepreneurs exceeds by far their specific areas of interest.

To understand better how this sort of 'something out of nothing' process is facilitated, we analyzed over 100 diverse cases of social entrepreneurs.[1] This lead us to hypothesize that they do this by building social capital; this not only helps them in pursuing their mission but also, as a side-effect, empowers the groups or communities in which they operate. Also we believed that social entrepreneurs have specific personality traits which facilitate this process (Praszkier 2007). These hypotheses were later positively verified in research (Praszkier et al. 2009).

However, there were two important research questions still left unanswered: what methods do social entrepreneurs use for building up and augmenting social capital? How does their activity influence the target population?

In this chapter we respond to those questions. In a portfolio of qualitative cases we studied the way social entrepreneurs build social capital and how this process affects the communities in which they operate.

We begin with an overview of the existing social change theories; from these we selected the Dynamical Social Psychology approach as the most appropriate for our purpose.

Social Change Theories

The process of change introduced by social entrepreneurs can be viewed from the perspective of various change theories.

Traditional Sociological Theories

The dynamics of change seems to resemble a movement through gradual stages, with each stage on a more 'advanced' level than the one before. This point of view

---

[1] They were all Fellows of the International Association of Social Entrepreneurs, Ashoka, Innovators for the Public, www.ashoka.org.

is reflected in evolutionary social change theories (Auguste Comte, Lewis Henry Morgan, Herbert Spencer).

Development through gradual stages often involves a potential conflict (for example some participants want to maintain the status quo, whereas the others want to implement novelties). Some sociologists (e.g. Lewis Coser) perceive conflict as a positive force for change. Constructive conflicts, where issues are publicly discussed and a solution found, build up and strengthen civic participation and democracy. (Other kinds of conflict, on the other hand, those based on hate and desire to rout the others, are destructive for individuals and the society.)

Social entrepreneurs perceive their target groups or societies as systems, similar to Henry Spencer's metaphor of society as the body whose proper functioning depends on its 'organs' working together. For example, Talcott Parson and William Ogburn, both functionalists, viewed societies as holistic and homeostatic entities in which cause-and-effect relationships are multiple and reciprocal. Social entrepreneurs focus both on functions and structures: usually their initial phase introduces new kinds of societal relationship, which enhances trust and cooperation and, as a result, galvanizes economic growth; in the next phase social entrepreneurs usually focus on modifying structures – procedures and the law – so as to enable the new idea to thrive and spread.

Social entrepreneurs understand, moreover, that societies consist of people and it is people's mindsets, beliefs, prejudices and attitudes that contribute to the way societies are organized, as well as vice versa. The psychological factors are essential in the process of change. In sociology, the representative of social-psychological approach was Max Weber, well known for his finding of an affinity between religions and directions of societal development. In the twentieth-century, sociologists (e.g., Everett E. Hagen, David McClelland) believed that in old, traditional societies personalities were generally uncreative and authoritarian, whereas in modern societies there exists an 'innovational personality.' The innovational personality persistently looks for new solutions and has a propensity for critically assessing common beliefs.

Kurt Lewin's Force-Field Theory

Kurt Lewin asserted that, prior to taking any action, one should analyze the factors (forces) that influence a given social situation (Lewin 2004). He viewed social processes as resulting from the interplay of social forces. Social systems are usually at the points of balance where every force is balanced by its counterforce. To achieve change one must first see the situation as a whole, then analyze the mosaic of social forces that aggregate to maintain the current balance of the social field, and finally draw out which forces are those driving toward the desired change and which would interfere (barriers or hindering forces). This 'Lewinian' way of thinking helps social entrepreneurs find the best point of application of force, enabling the slightest impulse to release maximum energy, leading to change.

Lewin also pointed out that change is frequently short-lived and that the situation soon returns to the previous level. So it is critical to include permanency of change in the planned objectives, and to equate permanent change with changing the equilibrium of the system. He asserted that to bring about any change, the balance between the forces which maintain the social self-regulation at a given level has to be upset through a process of unfreezing the present level, moving to the new level and freezing the situation on a new level. Since any level is determined by a force field, permanency implies that the new force field is made relatively secure against change.

Bubbles Theory: Social Change in the Perspective of the Dynamical Social Psychology

The bubbles theory derives from Dynamical Social Psychology. It is based on the dynamical theory of the attitude change (see this volume, Nowak et al., Chap. 1). It is this theory that forms the basis of our theory of social entrepreneurship.

Social Entrepreneurs as Change-Makers

What is a social entrepreneur? How can we recognize one when we see one? Change is recognized as an essential and indispensable factor for social entrepreneurship (Dees 1998; Drucker 1985; Martin and Osberg 2007; Roberts and Woods 2005), but several definitions have been given.

The definition of social entrepreneurship and the social entrepreneur provided by Dees (1998), claims that social entrepreneurs play the role of change agents in the social sector by:

- adopting a mission to create and sustain social value (not just private value);
- recognizing and relentlessly pursuing new opportunities to serve that mission;
- engaging in a process of continuous innovation, adaptation, and learning;
- acting boldly without being limited by resources currently at hand;
- exhibiting a heightened sense of accountability to the constituencies served and for the outcomes created.

Martin and Osberg (2007) see social entrepreneurs as those who:

- target underserved, neglected, or highly disadvantaged populations;
- aim at large-scale, transformational benefits that accrue either to a significant segment of society or to society at large.

For the research presented in this chapter we adopted the Ashoka, Innovators for the Public,[2] definition. Ashoka has been in the field of social entrepreneurship since 1980, and its definition is considered to be the most comprehensive (Bornstein 2004). The criteria for selecting Ashoka Fellows are (Drayton 2002, 2005; Hammonds 2005):

- having a new idea for solving a critical social problem;
- being creative;
- having an entrepreneurial personality;
- envisioning the broad social impact of the idea;
- possessing an unquestionable ethical fiber.

In this chapter, by social entrepreneurs we will understand Ashoka Fellows who were elected through a rigorous process[3] based on these criteria.

The Role of the Individual in Introducing Social Change

Dynamical models of social change stress the importance of the individual in the process of fostering change. Our computer simulations analyzing the diffusion of change revealed that the key role is played by leaders (Nowak et al. 1990), and specifically by social entrepreneurs (Praszkier and Nowak 2005). Social change in this case is generated internally in the system (Nowak 1996), social entrepreneurs fostering an endogenous kind of change process (Praszkier and Nowak 2011). The process involves mutual interactions between agents on the micro-scale, which leads to change processes on the macro-scale. This is one of the differences between social activists, pushing for change head-on, and social entrepreneurs, who play the role of change-catalysts and, through that process, as Peter Drucker said, they change the performance capacity of the society (Gendron 1966). They are doing so through creating a new texture of the social fiber: social networks and social bonds between people.

Research 1: Methods Used by Social Entrepreneurs

As a first step we wished to identify the specific methods used by social entrepreneurs, not only to build social capital, but also to achieve durable and irreversible results. Initially, we conjectured that they trigger a bottom-up social process followed by more concrete and detailed methods. Finally, we came up with several working conjectures concerning the methods that they use. Social entrepreneurs:

---

[2] See: www.ashoka.org.

[3] See: Ashoka selection criteria: www.ashoka.org/support/criteria.

- perceive the groups that they are working with as systems, seeing both the wholeness as well as the internal interrelations;
- have a high level of perceptiveness of people's and groups' needs, both visible as well as latent;
- pursue their social mission through a bottom-up process of empowering individuals and groups;
- enhance the propensity for cooperation within the group;
- bring and enhance trust and relationships based on this trust.

Those conjectures were supported in our qualitative pilot studies, drawing from the knowledge of the Ashoka experts participating in the selection process to Ashoka fellowship and from the case studies analyzed by qualified judges.

Evaluation of Social Entrepreneurs by Ashoka Experts Serving as Qualified Judges

The research group was all the 52 Polish Fellows elected at the time to Ashoka. They were evaluated by five qualified judges who were selected out of the pool of Ashoka experts who had served as panelists in the selection process to Ashoka fellowships. For each of the 52 Polish Fellows, the five judges completed a questionnaire with 24 questions such as:

- Does she/he have an ability to look at the target population in a systemic way, perceiving the wholeness and the complicated dynamical system?
- Is she/he sensitive to the (often latent) needs and potentials of the society?
- Does she/he make people feel empowered?
- Does she/he change competition into cooperation?
- Is she/he changing distrust into trust?

Evaluation of Interviews and Case Studies by Qualified Judges

This questionnaire, slightly modified, was again used to evaluate five case studies:

- Dorota Komornicka, empowering underserved rural and mountainous communities in southern Poland;
- Kazimierz Jaworski, developing the marginalized, south-east region of Poland;
- Krzysztof Czyżewski, cross-border and cross-cultural integration of various cultures on the border of Russia, Belarus and Lithuania;
- Krzysztof Stanowski, building civil society in ex-Soviet republics;
- Wasław Idziak, bringing jobs through empowering rural communities with a high unemployment rate in northern Poland.

Added to the questionnaire were several non-social-capital-building questions (relating, for example, to bringing experts from outside and patronizing the community) in order to evaluate not only if the social entrepreneurs are building social capital, but also if they are keen to use methods which are not building social capital.

All of those five cases encompassed a description, based on the Ashoka archives, and an in-depth interview, see Table 9.1 below, and were evaluated by three competent judges who were young-generation social activists.

Results

1. Social entrepreneurs are building social capital while pursuing their social mission (Praszkier et al. 2009).
2. Both qualitative studies confirmed the following methods used by social entrepreneurs (Praszkier 2007):

   - Social empathy:

     - Understanding (in many cases instinctively) the various types of potential (sometimes latent), embedded in groups and/or societies.
     - Understanding the pains, needs, frustrations of groups and/or society, as well as their hidden dreams and desires.
     - Being aware of the latent tendencies, i.e., in which directions groups or societies could potentially go.
     - Identifying areas in which groups/societies would eagerly cooperate, and where the likelihood of success is relatively high (the first success is critical for triggering the process of change).
     - Identifying areas of motivation (possibly latent) through responding to some important (maybe seemingly invisible) needs.

   - Empowering groups, avoiding disempowering methods:

     - Facilitating the change process in a way that empowers people and groups.
     - Ensuring that others fully experience the reality of their success.
     - Avoiding a patronizing, top-down teaching style; this means, for example, avoiding bringing experts from the outside as opposed to launching an internal educational mechanism that would trigger a process of mutual learning within the groups or communities.
     - Minimizing and gradually diminishing their own role so that the community could generate its own leaders

   - Modifying parameters:

     - Instead of confronting problems head-on, modifying parameters such as trust, optimism, hope, propensity for cooperation, and bringing groups/ societies to the point where it becomes natural to come together over a given idea.

Table 9.1 The analysis of archives, conversations with interviewees and the data assembled from field-observation

Social entrepreneur	The departure point preparing the ground for launching subject's idea	The existing projects
Dorota Komornicka (Community Fund for the Śnieżnik Mountain – Bystrzyca Kłodzka)	Informal meetings in interviewee's home, including culinary-entertainment weekends with neighbors and old friends from Warsaw; this evolved in 1976 into a foundation of the association Club ZDANIE, which played the role of a discussion club for dissidents (during the totalitarian regime till '89) and other interested participants Participation in several editions of the grant distributed by the Foundation of Local Democracy Development (including a study-trip to USA) Drawing from that experience, launching the Local Community Fund	International youth exchange between Germany, France, Czech Rep., Slovakia and Poland The scholarship program for children and youth, including educational stipends Grants for children and youth for social projects (e.g. cleaning the river banks, caring for abandoned graves, campaigning against throwing litter in public places) The development of agro-tourism Annual fetes, showing the result of the aforementioned children's and youth projects; also displaying the cross-generation projects Monitoring the river pollution Launching folk music assembles Issuing school newsletters Building sport centers Screening children for faulty posture (4k children were checked) Periodically screening of breast and prostate cancers Founding five rehab. centers (in Bystrzyca Kłodzka, Lądek Zdrój, Stronie Śląskie, Duszniki Zdrój i Kudowa Zdrój, with two more in process: Kłodzk and Międzylesie); currently 200 children participating
Ewa Smuk-Stratenwerth (Assocoation "Ziarno" – Grzybowo)	Spreading the idea of healthy food Campaign "Human Birth-giving"	Organizing Annual Earth Day Inviting folk music and dancing assembles from various countries

	Founding the Ziarno association (rewarded for actions for the rural communities)	Organizing fetes and village bazaars for the Grzybów region inhabitants
	Launching a rural education center Arka	English courses for children and youth
		Educational courses for inhabitants from villages as well as from cities, aimed at children, families, teachers and farmers
		Publishing manuals disseminating the education for people from farming areas
		Organizing workshops promoting organic farming and sustainability
		Publishing a local bulletin News from the Vistula River, co-edited by the inhabitants
		Organizing school programs adjusted to disabled or problem-children needs (annually there are over 2,000 children from all Poland participating, also school groups from Scandinavian and Baltic countries)
		Stipend programs for youth (since 1995)
Krzysztof Margol (The Nidzica Development Foundation NIDA – Nidzica)	In the 1990s was a mayor of the Nidzica county. At that time most of the employers dropped away which resulted in a high, 30 %, unemployment rate	Free consultation station, helping with processing the applications for grants for launching new ventures
	In 1994 resigned from the mayor's post and launched the Nida Foundation	Credit guarantee fund aimed at incubating new jobs, with the focus on employing the local jobless
	In 1995 succeeded to be one of the beneficiaries from the first edition of local initiatives program (Structural Development for Selected Regions)	The local community fund building an endowment for distributing jobs-generating microcredits
		Educational projects aimed at all generations, also at teachers; the focus was in English teaching

(continued)

Table 9.1 (continued)

Social entrepreneur	The departure point preparing the ground for launching subject's idea	The existing projects
Krzysztof Czyżewski (Borderland Foundation, Sejny)	Organizing in the 1980s an annual International Workshops of Alternative Culture in Czarna Dąbrówka cooperation with other cultural animators from "Gardzienice" and "Stop" theaters and with the Cultural Center "Dąbrówka" in Poznan; also with the local Cultural Center in Dabrowka Czarna in the Kaszuby region In 1994 founding the Borderland Foundation	Realizing a variety of artistic and educational program aimed at youth, for example: documenting the history of the region, education for tolerance, respect for tradition and revitalizing the vicinity. This was done through projects such as Building the House, where children and youth planned and designed a real building according to their tradition, along with sharing with others their stories explaining the cultural and religious meaning of particular elements
Wacław Idziak (The Koszalin Social-Cultural Association)	In the 1990s he worked in the Agency for Regional development in Koszalin Built strategies for local communities Launched the Koszaling Sport and Culture Association	Carried on the project of the first thematic village in Wierzbinek, currently turned into a Global Center Wierzbinek Realized the project of five thematic villages (within an EU project) Podgórki – the Fairy Tales and Play Village Sierakowo Sławieńskie – Hobbits Village Iwęcino – The End-of-the-World-Village Paproty, Mazes and Sources Village Dąbrowa – Healthy Life Village (In the period of May–December 2008 there were 16,000 tourists visiting those five villages) Moreover, the Thematic Villages also provide artistic and educational programs

Project "Thematic Village": a village with high unemployment was shaped into artisans' pot production historical style attracting tourists

- Identifying the best starting point:
  - The departure point is usually innovative, being the seed for the new attractor; it is constructed in a way that leads to the first success; the groups/societies experience and enjoy this first success and, as a result, are motivated to demonstrate further cooperation.
  - The departure point could be something other than the real goal, as the latter is often the area of a core conflict or a long-term struggle and resistance.

Research 2: The Impact on Groups and Societies

Method

There were two sources of data used to assess the impact of social entrepreneurs on groups and societies. The first source was from a questionnaire specially developed for this purpose. The second source was observations made in regions where social entrepreneurs operated, including interviews with local community members, three or four in each community.

Participants in the Questionnaire

One hundred and eight subjects participated in the questionnaire study. Their ages ranged from 17 to 63 (M = 42.05; SD = 10.43); there were 71 women (65.7 % of the sample) and 37 men (34.3 %).

The subjects were socially active individuals collaborating with five Polish social entrepreneurs: Ewa Smuk-Stratenwerth (Grzybowo; N = 13 subjects), Dorota Komornicka (Bystrzyca Kłodzka; N = 30), Wacław Idziak (Koszalin; N = 29), Krzysztof Czyżewski (Sejny; N = 20), and Krzysztof Margol (Nidzica; N = 16). It is worth mentioning that the 29 collaborators of Wacław Idziak lived in five 'thematic villages'[4] near Koszalin (Sianów county). A description of the projects in which these five entrepreneurs were involved is given in Table 9.1.

The majority of the sample (N = 60, 55.6 %) had academic education. A quarter (N = 27) had high school level education. The rest had vocational education (N = 14, 13 %) or elementary school level education (N = 2, 1.9 %). The subjects in the sample had been socially engaged for between 1 and 43 years (M = 11.59, SD = 8.96, Me = 10, Mo = 10). They were engaged in various projects: educational, cultural, sports, environmental, and health issues. 72 subjects were involved in one project, 15 in two projects, and 5 in three projects (16 no response; M = 1.27, SD = 0.55).

---

[4] A project sponsored by EQUAL.

Conjectures and Questionnaire

The conjectures were that the participants would be highly engaged, that they would predict their project's success and that they would report a high level connectivity between their collaborators during all the phases. The final conjecture was that the communities in which social entrepreneurs operate would have a highly developed social capital, using Putnam's (1993, 2000) definition. According to Putnam's interpretation, social capital facilitates high levels of trust and norms in societies; this enables a higher level of performance which, in a feedback loop, enhances trust. One successful undertaking builds connections and trust, which become social assets available also in project areas other than the current one. Individuals trust even more those who have already proved to be trustworthy. A positive feedback loop is thus created between trust and cooperation (Hardin 2001).

All subjects answered a questionnaire consisting of six questions, all but one on a three or four point scale – see Table 9.2. The questions asked for the subject's degree of engagement in current projects facilitated by social entrepreneurs, the nature of the emotional aspect of their engagement, their evaluation of whether the project would succeed or not (a yes-no question), and the frequency of contacts with other participants. The latter was probed over three phases of the project: planning, realizing and recapping of the project.

Analysis of the Questionnaire

The mean and standard deviation of the answers to all but one of the questions are shown in Table 9.2. The findings indicate a high level of involvement of the socially active subjects; additionally, subjects have positive or very positive emotions toward their projects. They maintain frequent connections with their collaborators at all the stages of their projects. For the remaining question, we asked for their assessment of the likelihood of their project's success: 66 individuals pointed out that the project that they were currently involved in would succeed, and only 2 that it would not. This indicates a high level of optimism concerning the realization of their projects.

Qualitative Analysis

Most of the 108 participants in the five communities were volunteers in the social projects, being employed elsewhere.[5] They not only linked their professional activity with social projects, but also influenced and engaged in social actions

---

[5] The research was carried on by Agata Zablocka-Bursa in 2008.

Table 9.2 Basic statistical information on answers to all but one of the questions

Scale	Possibility of answer	M	SD	Interpretation
Self-evaluation of engagement	1. Low 2. Middle 3. High	2.76	0.47	High
Emotional aspect of engagement in project	1. Very negative 2. Negative 3. Positive 4. very positive	3.47	0.59	Positive and very positive
Frequency of contacts with other participants in planning phase	1. Seldom (once in month and less) 2. Middle (two to three times in month 3. Often (at last once in week) 4. Very often (twice in week and often)	3.27	1	Often
Frequency of contacts with other participants in realizing phase	1. Seldom (once in month and less) 2. Middle (two to three times in month 3. Often (at last once in week) 4. very often (twice in week and often)	3.17	0.81	Often
Frequency of contacts with other participants in recapping phase	1. Seldom (once in month and less) 2. Middle (two to three times in month 3. Often (at last once in week) 4. Very often (twice in week and often)	3.03	0.96	Often

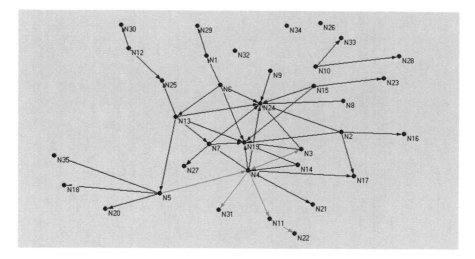

Fig. 9.1 Visualization of network analysis in Nidzica in which participants were asked: "With whom do you talk to on topics related to the social activity?" The links are between the participant asked and, at the arrow heads, the persons they mentioned. Colors on the links indicate the social activity (project)

within their employment settings: schools (various levels), preschools, police, fire-brigades, family care centers, libraries, forestry management, rehabilitation center, cultural units, Red Cross, borstal, municipal administration; also business firms, associations and foundations; finally, farms and agro-tourism centers. Involving their work environment indicates high level of commitment to the social activities.

In all the communities, interviewees (N = 17: 4 in Bystrzyca Kłodzka, 3 in Grzybowo, 4 in Koszalin, 3 in Sejny and 3 in Nidzica) pointed to similar 'small-steps' mechanisms which drew them into the activity carried on by the social entrepreneur. The first step was getting acquainted, informal conversation and the social entrepreneur listening to their problems. Next, there was a small event organized by the social entrepreneur, where the subject got more acquainted and mutual trust was aroused. The social entrepreneur's goal for those events was to assemble a small, well collaborating group, who could count on each other, feel mutual support and trust. These events were organized regularly. A cooperating network of people emerged as a result of these events. An example of such a network that was observed is shown in Fig. 9.1; note the two hubs – nodes with a large number of connections – N24 and N19. The third step was getting involved in working with the local community; at first with a small group, then gradually expanding the target to a broader population.

Those meetings and realizing the projects together, plus small success, enhanced the propensity for cooperation. Collaborators were a source of support, which was especially important in crisis situations. Individuals opened up and shared their ideas; also they began planning together. Various possibilities for raising new resources for new projects were considered; this led to the starting of new civic organizations focused on concrete programs.

In all those communities interviewees highlighted the significance of organizing training, workshops and lectures, the goal of which was not only to impart new knowledge, but also to maintain contacts; these are methods which are known to enhance social capital (Praszkier et al. 2009). Moreover, the training courses raised their self-reliance, as they were convinced that learning raised their value as potential employees.

The interviewees also pointed out that many of the social projects involved children and youth, for instance by providing some extra-curriculum activities, especially in the Bystrzyca Klodzka region; also in Nidzica, school students were included in planning some social projects. In Grzybow ecological workshops for youth were organized; in Sejny youth was involved in creating the Sejny Chronicles. Those projects often included cooperation between the generations. Such connections cover the cross-generational gaps were especially important for the oldest citizens who were experiencing marginalization. As a result, adults were proud of the accomplishment of the children of their region, which stimulated their support. The research revealed that jointly realized projects by children and adults bond the local community and increase trust and cooperation. Social bonds are reinforced by cyclical and long-term projects.

Interviewees mentioned the significance of projects raising the attractiveness of their region. Examples: 'Thematic villages' in Nidzica and Koszalin, the agro-tourism in Grzybow, or appealing events in Sejny and Bystrzyca Klodzka. It appeared that often neglected and under-financed regions started developing because the community found new inspiration and hope. People see growth opportunities through cooperation (for example cross-cultural cooperation in Sejny, or neighbor-to-neighbor cooperation in other regions). Social entrepreneurs' community projects are usually diverse, encompassing various elements, for example religion, education, tourism, sport, nutrition, or cultural events. Often those projects link entrepreneurship with promotion of the history and culture of the region.

One of the important factors pointed out by the interviewees was the significance of disseminating the information on their region and its activities. Such external presentations engenders pride in their own achievements and, as a consequence, increases their efficiency. Moreover, the social projects were all promoted in local media and in some cases in national or international media as well.

Interviewees indicated that the unquestionable upside of their social projects was that they were, on the one hand, cost-effective (at least in their initial phase), and, on the other hand, they engaged most of the inhabitants. Examples: micro-credit system in Nidzica, micro-stipends in Bystrzyca Klodzka, thematic villages near Nidzica and Koszalin, assembling historical materials in Sejny or cooking recipes in Grzybow.

The realization of common projects, even if small, together with small successes at various stages of these projects, causes people to better understand and concentrate on themselves, their families, neighbors, and the place they live in. People start having dreams; also they realize that they have some capacity to take action. They see that joint actions yield tangible results; they perceive that not everything depends on others and that one doesn't need to wait for external investments or institutions to bring change. At some point it appears that people learn to dream, to cooperate, to learn from each other; also that they start perceiving the place where they live as positive. New, previously unnoticed, perspectives appear. People stop waiting for external aide. They engage various local institutions, for example local government, civic organizations, schools or local business. On the grounds of these qualitative data, we believe that there was most probably spreading of 'bubbles of new' in the 'sea of old.'

Conclusions

This chapter was aimed at identifying the mechanisms of social change processes introduced by social entrepreneurs and the impact of those changes on the communities where they operate. The qualitative research confirmed the hypothesis that social entrepreneurs, along with pursuing their social mission, empower individuals, groups and societies through building trust and propensity for cooperation (social capital). They do this through the high level of social empathy that they have, i.e. through a deep understanding of the latent needs, dreams and drives of the groups and societies. They find a specific departure point for launching their projects, so that the project provides fast and tangible success and builds trust and cooperation; in a feedback loop, social capital has an impact on the permanence of the change process. Additionally, it empowers groups and societies and strengthens their coping abilities.

The research also demonstrated that social entrepreneurs engage the inhabitants in various social projects (educational, social, environmental, health, etc.). Those projects increase the quality of life. Moreover, they change the psychological characteristics of the participants by raising their feeling of being supported, as well as their level of optimism and trust toward others. Participants learn new skills, such as the knowledge of how to plan and write up a project, as well as the propensity for cooperation. Finally, their level of coping abilities and resourcefulness is enhanced.

Participating in the success of the project made interviewees satisfied with their own achievements, which raised their self-esteem. Many of the realized projects also contributed to making the region more attractive, which augmented local pride.

One of the visible results of social entrepreneurs' activities is that a higher level of cooperation and communication between projects' participants was achieved.

Concluding, social entrepreneurs launch change processes which are having a multi-pronged and permanent impact on both the individual and the societal levels. There appear 'bubbles of new' which are characterized by a high level of trust and cooperation (i.e. social capital in Putnam's understanding).

References

Bornstein, D.: How to Change the World: Social Entrepreneurs and the Power of New Ideas. Oxford University Press, New York (2004)

Dees, J.G.: The Meaning of social entrepreneurship. Draft report for the Kauffman Center for Entrepreneurial Leadership, p. 6. http://www.redalmarza.cl/ing/pdf/TheMeaningofSocialEntrepreneurship.pdf (1998). Retrieved 1 Jan 2011

Drayton, W.: The citizen sector: becoming as entrepreneurial and competitive as business. Calif. Manage. Rev. **44**(3), 120–132 (2002)

Drayton, W.: Where the real power lies. Alliance **10**(1), 29–30 (2005)

Drucker, P.F.: Innovation and Entrepreneurship. Harper Business, New York (1985)

Gendron, G.: Flashes of genius. Interview with Peter Drucker. Inc. Mag. **18**(7), 30–39. http://www.inc.com/magazine/19960515/2083.html (1966). Retrieved 1 Jan 2011

Hammonds, K.H.: A Lever Long Enough to Move the World, Fast Company. http://www.fastcompany.com/magazine/90/open_ashoka.html (2005). Retrieved 11 Nov 2010

Hardin, R.: Conceptions and explanations of trust. In: Cook, K.S. (ed.) Trust in Society. Russell Sage Foundation, New York (2001)

Lewin, K.: Resolving Social Conflicts: Field Theory in Social Science. American Psychological Association, Washington, DC (2004)

Martin, R.L., Osberg, S.: Social entrepreneurship: the case for definition. Stanf. Soc. Innov. Rev. (Spring), 29–39 (2007)

McLeod, J., Thomson, R.: Researching Social Change: Qualitative Approaches. Sage Publications, London (2009)

Nowak, A.: Bąble nowego w morzu starego: podwójna rzeczywistość okresu przemian społecznych. In: Marody, M., Gucwa-Leśny, E. (eds.) Podstawy życia społecznego w Polsce, pp. 229–251. Instytut Studiów Społecznych, Uniwersytet Warszawski, Warszawa (1996)

Nowak, A., Szamrej, J., Latané, B.: From private attitude to public opinion: a dynamic theory of social impact. Psychol. Rev. **97**(3), 362–376 (1990). doi:10.1037/0033-295X.97.3.362

Praszkier, R.: Methods used by social entrepreneurs for introducing social change. Unpublished doctoral dissertation, Psychology Department, University of Warsaw, Warsaw (2007)

Praszkier, R., Nowak, A.: Zmiany społeczne powstałe pod wpływem działalności przedsiębiorców społecznych (Social changes triggered by social entrepreneurs). III Sect. **2**, 140–157 (2005)

Praszkier, R., Nowak, A.: Social Entrepreneurship: Theory and Practice. Cambridge University Press, New York (2011) (in process)

Praszkier, R., Nowak, A., Zablocka-Bursa, A.: Social capital built by social entrepreneurs and the specific personality traits that facilitate the process. Psychol. Spol. **4**(1–2), 42–54 (2009)

Putnam, R. D.: The Prosperous Community: Social Capital and Public Life. The American Prospect, **13**: 35–42 (1993)

Putnam, R. D.: Bowling alone. The collapse and revival of American community. Simon and Shuster, New York (2000)

Roberts, D., Woods, C.: Changing the world on a shoestring: the concept of social entrepreneurship. Univ. Auckl. Bus. Rev. (Autumn), 45–51. http://www.uabr.auckland.ac.nz/files/articles/Volume11/v11i1-asd.pdf (2005). Retrieved 11 Nov 2010

Chapter 10
Interpersonal Fluency: Toward a Model of Coordination and Affect in Social Relations

Wojciech Kulesza, Robin R. Vallacher, and Andrzej Nowak

Abstract This chapter investigates the role of interpersonal coordination in promoting positive affect in social interaction. We propose that coordination in the successful pursuit of a goal increases positive affect which in turn strengthens those social ties for maintaining relationships that are characteristic of effective coordination. In two experiments, participants performed a virtual task on a computer in tandem with a virtual co-actor, whose behavior was generated by a computer program. In Experiment 1, successful performance on a complex task required behavior coordination. In Experiment 2, behavioral coordination was manipulated independently of successful performance. Results showed that behavioral coordination promotes liking (Experiment 1) and that this effect is manifest under success but not failure (Experiment 2). Discussion centers on the adaptive value of behavioral and emotional coordination, the link between coordination and other psychological processes, and the conditions that promote rather than hinder the attainment of such interaction fluency.

The research was supported in part by grant 1 H01F 035 27, NN106 220138 and a Mobility Plus grant 644/MOB/2011/0, from the Polish Ministry of Science and Higher Education, and a grant from the Future and Emerging Technologies programme FP7-COSI-ICT of the European Commission through project ICT Collectives. We thank Wieslaw Bartkowski for his help in writing the computer programs employed in Experiment 2.

W. Kulesza (✉)
Warsaw School of Social Sciences and Humanities, Chodakowska 19/31, Warsaw 03-815, Poland
e-mail: wkulesza@swps.edu.pl.

R.R. Vallacher
Department of Psychology, Florida Atlantic University, Glades Rd. 777, Boca Raton, FL 33431, USA
e-mail: vallacher@fau.edu

A. Nowak
Department of Psychology, University of Warsaw, Stawki 5/7, 00183 Warsaw, Poland
e-mail: andrzejn232@gmail.com

A. Nowak et al. (eds.), *Complex Human Dynamics*, Understanding Complex Systems,
DOI 10.1007/978-3-642-31436-0_10, © Springer-Verlag Berlin Heidelberg 2013

Establishing social relations is a preoccupation for most people. The motivation for doing so, however, goes beyond fulfilling a desire for affiliation or belongingness per se. For people to accomplish a large proportion of their goals, they need to coordinate their actions with others in dyads or groups. The quality of coordination is often decisive in the attainment of n-person goals, whether in sports, families, corporations, work teams, or various ad hoc situations calling for people to perform a common task. We propose that positive affect (e.g., liking) among individuals signifies that coordination in the pursuit of goals has been achieved and also strengthens social ties so as to maintain the relationship that produced the success-ful coordination. Positive affect, in other words, both signals and supports the efficient performance of well-functioning dyads and groups. By the same token, negative affect among individuals both leads to poor coordination and disrupts the effectiveness of dyads and groups. In this chapter, we develop the rationale for the role of coordination in social relations and we present two experiments designed to highlight critical features of the proposed link between coordination and relationship formation, goal attainment, and affect.

Broadly defined, coordination refers to a harmonious relationship among the elements comprising a system that evolves in time (cf. Jirsa and Kelso 2004; Kelso 1995; Nowak et al. 2002; Turvey 1990). This basic depiction finds expression in a host of psychological processes. In perception, coordination is manifest as the efficient processing of elements in a stimulus array to generate a coherent and stable pattern (e.g., Lewenstein and Nowak 1989; Winkielman and Cacioppo 2001; Winkielman et al. 2002; Zajonc 1980). Such perceptual fluency has a counterpart in cognition and social judgment: distinct cognitive and affective elements in the stream of consciousness support one another in forming higher-level mental representations (cf. Tesser 1978; Vallacher et al. 1994). Both perceptual and cognitive fluency, meanwhile, are associated with positive affect toward the stimu-lus (e.g., Winkielman and Cacioppo 2001; Winkielman et al. 2002; Zajonc 1968, 1980). Negative affect regarding a stimulus, on the other hand, is associated with differentiated thoughts and judgments (e.g., Coovert and Reeder 1990; Isen 1990; Kanouse and Hanson 1971; Pratto and John 1991).

The importance of coordination has also been established for the mental repre-sentation and control of any action by an individual. A person can identify any action and enact it at different levels of integration between its components, from its lower-level elements to its higher-level meanings and implications (Vallacher and Wegner 1987). With increasing action mastery, there is a tendency for the lower-level elements to become progressively coordinated with respect to the action's higher-level representations. Efficient and effective action performance, then, represents a well-orchestrated interplay of action components over time, allowing the actor's attention to focus on the larger goal or implication of what they are doing. An action performed in this manner is associated with 'flow,' a positive affective state that supports the action's performance and pushes the action to a yet higher-level of coordination between the components of the action (cf. Csikzentmihalyi 1990). Even an action that is mastered and thus identified in high-level terms can, however, be disrupted, thereby reinstating mental representations of the action's lower-level elements. When an action is disassembled in line with this scenario, there is a corresponding increase in negative affect on the part of the actor (cf. Nowak and Vallacher 1998; Simon 1967; Vallacher 1993). In line with our rationale, then, well-coordinated and efficient action increases positive affect which in turn perpetuates such action, whereas negative affect signals poorly coordinated action and promotes further disruption.

Extending this rationale to dyadic and group functioning suggests that social relations can be framed in terms of interpersonal fluency. When the members of a dyad or group achieve high coordination with one another, they can function with correspondingly greater efficiency and effectiveness on a common task. The attainment of functional coordination elicits positive affect, which in turn strengthens the social ties so that the dyad or group will be maintained in the future (cf. Nowak and Vallacher 1998; Nowak et al. 2002; Wiltermuth and Heath 2009). Conversely, negative affect signals poor coordination and functions to weaken social ties, decreasing the likelihood that the dyad or group will reassemble in the future.

The role of coordination in social interaction and social relationships has received considerable theoretical and empirical attention in recent years.[1] For the most part, research on this topic has focused on simple forms of coordination, such as mimicry of motoric actions and bodily postures. Mimicry, for example, has been shown to increase perceptions of interpersonal similarity (Abele and Stasser 2008), create feelings of closeness (Ashton-James et al. 2007; van Baaren et al. 2003), create shared preferences (Tanner et al. 2007), and increase the likelihood of offering assistance (van Baaren et al. 2004). Behavioral mimicry is easily and automatically achieved with almost anyone (e.g., Bargh et al. 1996; Chartrand and Bargh 1999; Kawakami et al. 2002; Meltzoff and Moore 1977; Strack et al. 1988; Strogatz 2003). Indeed, because of the automaticity associated with mirroring, mimicry, and interaction synchrony, control processes are necessary to prevent these basic forms of coordination from occurring.

Coordination can, of course, take far more complex forms (cf. Jirsa and Kelso 2004; Nowak and Vallacher 1998; Turvey 1990), and such forms are less likely to occur automatically, at least initially, and cannot be achieved by everyone. Members of a work group, for example, may experience difficulty in achieving effective temporal synchronization of their respective actions in pursuit of the group task. Even in a casual conversation, two individuals may experience awkwardness and disruption in their attempt to take turns talking and listening if they have different temperaments or are experiencing different moods (Nowak et al. 2002). So although there is an inherent press toward coordination in social relations, analogous to the self-organization characteristic of dynamical systems in other areas of science (cf. Strogatz 2003; Nowak and Vallacher 1998), interpersonal

---

[1] For example: Baron et al. (1994), Bernieri et al. (1988), Condon (1982), Condon and Ogston (1966), Decéty and Chaminade (2003), Dijksterhuis and Bargh (2001), Harakeh et al. (2007), Kendon (1970), LaFrance (1979), Lankin et al. (2003), Marsh et al. (2006), Newtson (1994), Nowak et al. (2002), Schmidt et al. (1990), Shockley et al. (2003), Wilson and Knoblich (2005).

fluency may be difficult or impossible to achieve between some people or for tasks that require complex forms of mutual dependency in peoples' actions over time (e.g., Nowak and Vallacher 2005; Vallacher et al. 2002). Computer simulations of interacting systems have revealed that similarity of the internal parameters of these systems (e.g. moods, action plans or temperament) is critical for the systems to achieve behavioral coordination (Kaneko 1993; Zochowski and Liebovitch 1997 cf. Nowak and Vallacher 2003, 2005; Nowak et al. 2002; Nowak and Vallacher 1998). If the respective parameters of the interacting systems are very different, synchronization may be impossible to achieve. Conversely, even when the two systems are weakly coupled (i.e., they have little direct influence on one another), similarity of their respective internal states promotes a similar temporal pattern of behavior. This finding suggests that similarity in internal states may be of critical importance in deciding with whom people attempt to achieve coordination (Hatfield et al. 1994; Simpson et al. 2003), at least under conditions of equal mutual influence. Positive affect following an interaction can thus indicate that coordination with this particular partner was achieved, and thus is likely to occur in future interactions.

This perspective goes beyond the assumption that mimicry (e.g., the chameleon effect) of any kind promotes liking between people (cf. Chartrand and Bargh 1999; Marsh et al. 2006). In unstructured interactions, the implicit goal of each person is to react to the other with the possibility of forming a relationship. We suggest that the tendency for mimicry to promote liking reflects coordination on this implicit goal rather than being an automatic effect of mimicry per se. There is evidence, for example, that people mimic others whom they find physically attractive (van Leeuwen et al. 2009), presumably because they desire to form a relationship with them. Increased liking, in other words, is the result of coordination on aspects that are relevant to the task of the dyad or group (Chartrand and Bargh 1999, Experiment 2). Coordination on irrelevant dimensions, in contrast, is unlikely to produced positive affect and bonding. Imagine, for example, two individuals talking about mutually important issues over a bonfire in a mosquito-infected area. Coordination in reacting to each other's comments, tone of voice, and bodily gestures will create the effect of rapport and liking. Their coordination in slapping mosquitoes on their own individual bodies, which doesn't represent a common task of the individuals, will not affect mutual liking. However, if the two individuals want to kill mosquitoes in their tent before going to sleep, their coordination in doing so is more likely to increase the bond between them than was the coordination in their conversation.

To establish the importance of the functional value of coordination, it is necessary to link coordination explicitly with successful and unsuccessful task performance. If, as the present rationale suggests, coordination produces positive affect because coordination promotes efficient and effective action, success versus failure should moderate the effect of coordination on interpersonal affect. Under success, coordination should enhance liking, in line with the research on behavioral synchrony and behavioral mimicry. Under failure, however, coordination should have no effect on liking or perhaps promote negative affect, so as to weaken the social ties and prevent further ineffective group action.

Overview of the Present Research

To provide evidence in support of our rationale, two experiments were conducted. In each experiment, participants were asked to perform individually a virtual task on a computer in tandem with what they believed was another person, but whose behavior was in fact generated by a computer program. In Experiment 1,[2] participants attempted to navigate a complex labyrinth, displayed on the computer monitor; their success was dependent on the degree of synchronization of the behavior of the simulated co-actor. In Experiment 2, participants attempted to virtually catch a sequence of falling balls while maintaining balance on a seesaw displayed on the monitor. A participant's success in this task was dependent on how well the simulated partner compensated for the participant's movements by its own movements on the virtual seesaw, and vice-versa. In both cases, successful coordination went beyond simple mirroring or mimicry, reflecting instead compensatory movements that enabled the dyad to complete the task in question.

Upon completion of the task in each experiment, participants answered a series of questions designed to tap their feelings about the partner, their degree of coordination with the partner, their feeling of task accomplishment, and their desire to work with the partner on future tasks. We predicted that successful coordination in both experiments would be associated with positive feelings about the partner. In Experiment 1, we expected coordination to have a positive effect on affect only if the coordination occurred on a dimension that was directly relevant to performance on the task, while coordination on any other dimension would be irrelevant to liking. In Experiment 2, we introduced a manipulation on performance feedback. We expected the effect of coordination to be moderated by performance feedback (operationalized as a comparison with other purported dyads). Specifically, coordination was expected to promote positive affect under success feedback but not under failure feedback.

Experiment 1

Method

Participants

Sixty women and twenty men were recruited from Psychology classes at Warsaw School of Social Sciences and Humanities in exchange for course credit.

---

[2] Originally presented in: Kulesza and Nowak (2003). Here we have recalculated gathered data and recapitulated findings.

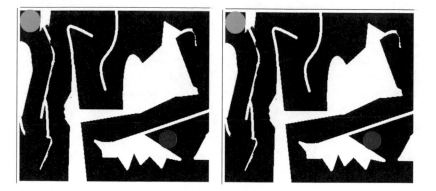

Fig. 10.1 Virtual labyrinth displays employed in Experiment 1

Procedure

Participants were informed that they would interacted individually with a second person (in fact simulated by a computer program) in the context of a computer game. The scenario for this game was the navigation of a virtual spaceship around stationary obstacles in a simple labyrinth with a single solution. The goal was for both spaceships to reach the end of the labyrinth without hitting the obstacles. The task was simply to navigate the spaceships through the labyrinth, with no explicit instructions concerning the time required or the number of obstacles avoided or hit.

The participant (person A) was presented with two displays of the same labyrinth on a computer monitor, with one portraying their spaceship and the other portraying the spaceship of the virtual partner (person B). Figure 10.1 shows how the two labyrinth displays were presented to the participants.

The cover story was that B could not see the obstacles due to damaged radar on his or her spaceship, so that the only way B could successfully navigate the labyrinth was by observation of the movement of A's spaceship. Successful completion of the game depended on both A and B reaching the end of the labyrinth. A was also told both spaceships may come under attack if detected by a mutual enemy. To avoid detection, A and B needed to change the color of their spaceships (by pushing a button) at least once every 15 s (but nothing happened if partners failed to change the color of their spaceships; all participants pushed a button at least once every 15 s. As A believed that B could see the color of its own spaceship, there was no need for coordination on color change.

The computer program provided two levels of coordination for both movement and color change independently. Under high coordination, B followed the movement of A with a slight time delay and a small amount of error to make the coordination realistic. Under low coordination, both the time delay and the error of movement were larger, although the error was sufficiently small that the task could be completed by all participants. With respect to change in color, the coordination manipulation was only on time delay not on error.

Upon completion of the labyrinth task, participants completed a questionnaire designed to assess their attitude toward the virtual partner on several dimensions, including trust, liking, and desire for interaction generally and with respect to a similar task in the future.

Dependent Measures

The questionnaire consisted of seven questions, each answered on a five-point scale ("fully disagree" to "fully agree"). Four questions assessed liking for the partner: (1) Do you like the other person? (2) Do you find the other person interesting? (3) Do you think that the other person is a nice person? and (4) Would you like to meet that person in the future? Three questions assessed participants' feelings about how well they coordinated with the virtual partner on the task: (1) Would you like to cooperate with the other person in the future? (2) Do you think that both of you would create a good team? and (3) Would you trust that person if you would have to cooperate in the future on a similar/another task?

Hypotheses

As movement coordination was relevant to task completion, we expected participants to express more positive affect for their virtual partner and higher satisfaction with the dyad's performance under high movement coordination than under low movement coordination. Because, however, the coordination of color change was not relevant to task completion we expected that high versus low coordination on this feature would not affect participants' affect toward the partner or their performance satisfaction.

Results

Preliminary Analyses

It is important to distinguish between affect per se and perceptions of effective coordination with respect to a task. Accordingly, the four questions tapping liking for the partner were analyzed for internal consistency to determine whether they collectively formed a reliable scale. Analysis shows that the items were highly inter-correlated, average $r(80) = 0.45$, $p < 0.001$, and thus formed a highly reliable scale (alpha $= 0.71$) which we labeled liking. The same approach was taken for the three items that assessed perception of task coordination. Analysis shows that the three items were highly inter-correlated, $r(80) = 0.54$, $p < 0.001$, and thus formed a highly reliable scale (alpha $= 0.85$), which we labeled team desirability.

Tests of Hypotheses

Because coordination with respect to movement was directly relevant to task completion, we predicted that participants in the high coordination condition would express more positive affect toward the other person than would participants in the low coordination condition. Results confirmed this expectation. There was a reliable main effect of movement coordination for liking, $F(1, 76) = 5.27$, $p < 0.05$, with greater liking in the high than in the low coordination condition (M = 4.5 vs. 4.1). This effect was also observed for team desirability, $F(1, 76) = 10.6$, $p < 0.002$. Participants in the high coordination condition felt more positive about their teamwork than did those in the low coordination condition (M = 3.8 vs. 3.26).

The virtual partner also coordinated its change of color with that of the participant, although it could have acted independently since coordination of color change was not directly relevant to task completion. Hence, coordination of color change was not expected to impact participants' affect and performance ratings. In line with this reasoning, analysis of the results showed that the degree of color coordination had no effect, neither on liking nor on team desirability, $F(1, 76) < 1$, not significant (ns) in both cases.

There was variability in the amount of time taken for task completion (M = 52.7 s, SD = 20.8 s). However, neither movement nor color coordination had a reliable effect on this variable, $F < 1$, ns in both cases.

Discussion

The results of this study provide support for the hypothesized role of coordination in promoting positive attitudes toward one's task partner. For both liking and perception of team desirability, participants who experienced high coordination with respect to a task-relevant behavior (i.e., movement) felt more positive about their partner than did those who experienced low coordination. However, with respect to a task-irrelevant behavior (i.e., color change), high coordination did not affect participants' feelings and perceptions of team desirability. This suggests that people do not value coordination for coordination's sake, but rather because of its potential instrumental value in achieving a group task.

There are two important caveats regarding these conclusions, however. First, coordination was limited to simple mimicry, such that the virtual partner repeated the movements of the participant. Clearly, coordination can take other forms. Indeed, anti-phase coordination – in which people must alternate their behaviors in time rather than display their behaviors at the same time – is ubiquitous in dyadic (and group) action. When two people talk, for example, effective coordination is defined in terms of turn taking rather than simultaneity. Experiment 1 did not address the role of this form of coordination in generating positive attitudes.

Second, in many task contexts there is feedback regarding the relative success versus failure of the behavior, and such feedback may promote attitudes toward one another that are independent of, and conceivably stronger than, the effects of coordination. In principle, two people could succeed without coordinating, or coordinate without achieving success. If coordination promotes liking because of its presumed facilitating role in goal attainment, performance feedback should moderate the effect of high versus low coordination. Specifically, the link between coordination and liking should be most apparent when there is explicit indication that coordination was associated with successful goal attainment.

The third caveat concerns the manipulation of coordination in this study. One could argue that when the virtual partner followed the behavior of the participant, this constituted successful social influence on the part of the participant. Conceivably, it was the perception of influence, not that of coordination, that promoted the attitudinal effects observed in this study.

Experiment 2

With these caveats in mind, a second experiment was performed. The task was again in the form of a computer game, but unlike the task employed in the first experiment it required anti-phase coordination. Participants individually interacted in a computer game with a virtual partner, whose behavior was simulated by a program that coordinated with the participant to either a high or a low degree. Upon completion of the task, participants received feedback on their performance, being told that they had performed relatively well (success) or relatively poorly (failure). As in Experiment 1, we hypothesized that a higher degree of coordination would result in a higher evaluation of the virtual person and of the team. This effect, however, was expected to be stronger when participants were provided success as opposed to failure feedback.

Method

Participants

Eighty undergraduates (60 women, 20 men) at Warsaw School of Social Sciences and Humanities participated as part of their course requirements.

Procedure

Each participant was seated individually in front of a computer running under a Windows operating system. They were able to control the position of a virtual basket located on the left side of a seesaw presented on the screen (see Fig. 10.2). Their task was to position the basket so that balls descending from the top of the computer screen would drop into the basket. Participants controlled the basket

Fig. 10.2 Virtual seesaw displays employed in Experiment 2

by using the 'O' and 'P' keys on the keyboard to move the basket to the left and right, respectively. They were informed that another person in the adjoining room would also be attempting to maintain their balance on the other side of the same seesaw while catching the balls in their basket. The behavior of the ostensible other person was in fact generated by a computer program. The shared goal of the participant and the virtual partner was to maintain the balance of both players, so that both could catch the balls without falling off the seesaw. Participants were told that the task would last 3 min.

The balls dropped from the top of screen at an average rate of 1 ball every 2½ s (for an average of 65 balls during the 3 min game), but with large variation in frequency so that participants could not anticipate precisely when they would have to make an adjustment in their position on the see-saw. The balls fell at all possible positions on the seesaw chosen randomly. If a ball missed the baskets, it would continue past the seesaw and disappear at the bottom of the screen. If the participant or the other person lost balance and fell off the seesaw, an overhead view of the seesaw would be displayed for a few seconds along with a message saying "wait" that would last for 3 s. The seesaw would then be displayed in its functional form again, with the participant and the other person on their respective ends, and the game would continue.

Throughout the game, the cumulative score (i.e., number of balls caught in both baskets) was displayed at the top of the screen, as were the individual scores of the players. The scores for the participant and the partner were fairly equal in all cases, and the cumulative scores also showed relatively little variability between teams.

Pilot Study

To calibrate the behavior of the other 'person' in this task, a pilot study was performed with 15 participants who played the game with versions of the computer program that differed with respect to the time delay in responding to a falling ball and the acceleration in movement in getting into an appropriate position to catch a ball. Together these parameters dictated the movement pattern and the success rate. Through successive iterations of variations in these parameters, a setting for each parameter was established so that the behavior and success rate of the virtual person matched the average behavior of the pilot participants. This reduced the likelihood that the participant and the virtual person would differ in their respective contributions to the cumulative score. These settings also ensured that the movements of the virtual person looked realistic and natural. None of the participants in the actual study voiced suspicion that the virtual person was not another naïve participant.

Independent Variables

Coordination. In all cases, coordination between the players was anti-phase. If one player moved in one direction (e.g., to the left) on the seesaw to catch a ball, the other player would have to move in the opposite direction (to the right) to maintain balance on the seesaw. The turning movement of the seesaw responded to the relative positions of the players with some delay attributable to the energy required to build momentum, so that if a player moved rapidly to catch a ball and returned to the starting position, the seesaw would remain stationary and thus not require compensatory movement by the other player.

The manipulation of coordination involved only the degree to which the virtual player (i.e., computer program) compensated for the movements of the participant when the ball fell on the participant's side of the seesaw. In the high coordination condition, 80 % of the program's movements were an appropriate compensatory response to the movement of the participant. In the low coordination condition, 22 % of the program's movements were an appropriate compensatory response to the participant's movements. In both conditions, when the ball fell on the program's side of the seesaw, the virtual person moved to catch the ball regardless of the participant's movements. It was up to the participant in this case to make the appropriate compensatory response so that both players could maintain balance on the seesaw.

Success-Failure Feedback. The running cumulative score, as well as the respective scores of the players, were displayed at the top of the screen. There was not an objective standard for performance in this novel task, so the cumulative score by itself did not signify good versus bad performance. To convey such feedback, upon completion of the task the experimenter informed the player of the cumulative score for their dyad and randomly indicated, on a 50–50 basis, independently of the coordination manipulation, that this score was either "good, well above the average performance of other participants in this study" (success) or "poor, below the average performance of other participants in this study" (failure).

Dependent Measures

Upon receiving the success versus failure feedback, participants were asked to complete a paper-and-pencil questionnaire tapping their feeling about having collaborated with the other person (one question), their assessment whether they would collaborate effectively in the future (one question), their desire to collaborate in the future with the other person (one question), their liking for the person in general (four questions, three of which were adapted from a question set developed by Wojciszke 2003), and their feeling that the other person understood their feelings (one question). Each item was answered on a five-point scale ("fully disagree" to "fully agree"). Based on an *a priori* distinction between feelings about the task collaboration and feelings about the other person, we created two scales: task-based liking (two items) and general liking (four items). Both scales had very high internal consistency (alpha = 0.77 for task-based liking and 0.87 for general liking). We also analyzed the single item assessing whether the other person had emotional understanding.

Results

Tests of Hypotheses

Analyses of variance were performed for the three self-report scales as a function of high versus low coordination and success versus failure feedback. For feedback, a reliable effect was observed only on task-based liking, $F(1,76) = 8.45, p < 0.05$, such that a greater sense of team compatibility was expressed under success feedback (M = 3.8 vs. 3.2). No significant effects were observed of success versus failure feedback on the other dependent measures. Thus, success was recognized by participants and affected their sense that they and the other person functioned well together and probably would again in the future. However, the success versus failure of their collaboration did not generalize to feelings of liking and emotional understanding.

Analysis of the results revealed a reliable main effect of the coordination manipulation on all scales: general liking, $F(1,76) = 4.81, p < 0.05$ (M = 3.73 vs. 3.33); task-based liking, $F(1,76) = 5.29, p < 0.05.05$ (M = 3.74 vs. 3.26); and emotional understanding, $F(1,76) = 4.64, p < 0.05$ (M = 3.3 vs. 2.8). Unlike feedback, the impact of coordination was not limited to perceptions of task effectiveness, but was also apparent in participants' affective response to the virtual person. Coordination, in other words, subsumed the effect of success per se, but generalized to positive impressions of the virtual person.

Analysis of the results also revealed a reliable interaction of coordination and success-failure feedback effect, both on general liking, $F(1,76) = 7.84, p < 0.05$, and on emotional understanding, $F(1,76) = 7.84, p < 0.05$, but not for task-based liking, $F(1,76) = 1.88$, ns. Simple effects analyses were performed to compare the effect of high versus low coordination within each feedback condition

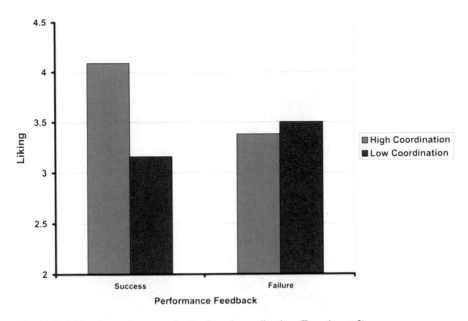

Fig. 10.3 Liking by performance feedback and coordination (Experiment 2)

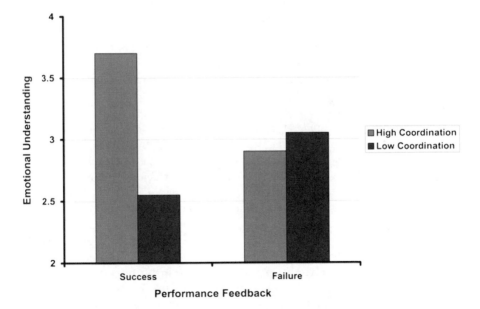

Fig. 10.4 Inferred emotional understanding by performance feedback and coordination (Experiment 2)

(see Figs. 10.3 and 10.4). Under failure feedback, coordination did not affect participants' liking for the other person, either in a general sense, t(38) = 0.41, ns, nor with respect to emotional understanding, t(38) = 0.45, ns. Under success feedback, however, analysis revealed a reliable simple main effect of coordination for both liking, t(38) = 3.66, p < 0.001, and emotional understanding, t(38) = 3.49, p < 0.001. Compared to low coordination, high coordination promoted greater liking (M = 4.09 vs. 3.16) and a stronger sense of emotional understanding (M = 3.7 vs. 2.55). High coordination, in other words, did not guarantee positive feelings on the part of participants. Coordination is valued because of its predictive value for effective performance, so it had a greater impact on participants' feelings when there was in fact evidence that it enhanced the team's effectiveness.

Discussion

The results of this study establish that the effect of coordination on liking is not restricted to in-phase synchrony (e.g., mimicry or the chameleon effect). Even when attempting to perform a task that requires attention, the ability to coordinate with someone in anti-phase promotes positive affect toward the partner. Moreover, heightened positive affect is not limited to preference as a task partner, but rather generalizes to liking for the person generally. In contrast, success and failure has a more limited effect on people's affect towards one another. As one might expect, the experience of success enhanced liking for the partner, but this effect was restricted to feelings about the person as a partner per se. Taken together, these results suggest that coordination itself is a more fundamental source of interpersonal affect than is the outcome of a common action.

The results of this study go beyond those of Experiment 1 by showing that the impact of coordination is moderated by the dyad's success versus failure on a task. Under success, high coordination promoted greater liking and a stronger inference of emotional understanding than did low coordination. Among 'unsuccessful' dyads, in contrast, liking and inferred emotional understanding did not differ for those who were well versus poorly coordinated. Stated differently, these results suggest that although success versus failure impacts one's desirability to continue working on similar tasks, it only transfers to liking and emotional understanding if the person feels sufficiently coordinated to think of the dyad as a team. People clearly value success, but the effect of success on liking is enhanced when it is perceived to be a result of effective coordination, presumably because this suggests that interaction with the partner is likely to promote future success on other tasks. In the absence of coordination, each person's sense of success versus failure is felt individually and does not affect feelings for the other person. However, if the person feels as though they have formed a functional team with the other person, the success versus failure is experienced in a correspondingly team-like manner. Success, then, is not limited to feelings about oneself but rather projects into a shared positive experience. Failure, by the same token, reflects on the team rather than simply on oneself.

General Discussion

Two experiments reflecting different paradigms and operational definitions point to the same basic conclusion: coordination is an important basis for liking. In Experiment 1, participants whose movements in navigating a labyrinth in virtual space were closely followed by the movements of a virtual partner expressed greater liking for the partner than did participants whose movements were not closely followed. In Experiment 2, participants who could perform a task successfully while coordinating in anti-phase to maintain balance on a seesaw with a virtual partner tended to like their virtual partner more than did participants who could not maintain balance while performing the task. So whether the issue is joint behavior in pursuit of a common goal or separate behavior in the context of coordination on a task, the results of these experiments suggest that being able to coordinate one's behavior over time with someone else sets the stage for positive feelings toward that person.

The results of these studies build upon earlier research demonstrating a proclivity for behavioral mimicry (e.g., Chartrand and Bargh 1999; van Baaren et al. 2004). That line of research demonstrates that people spontaneously adopt the motoric features of a person they observe, and that such coordination both facilitates, and is facilitated by, liking for that person. Such mimicry apparently occurs without conscious awareness (e.g., Lankin et al. 2003). Indeed, the potential for mimicry has been observed in newborn infants (e.g., Meltzoff and Moore 1977) and thus may represent a very basic feature of human relations.

The present research, however, goes beyond the research on mimicry in two ways, each of which is consistent with the presumed adaptive value of interaction fluency. First, the relevance of coordination to affect in social relations may entail more than in-phase synchronization such as that observed in research on behavioral mimicry. In Experiment 2, coordination involved anti-phase movement with another person in order to maintain balance while each attempted to succeed at a task (catching balls). In real-world settings, of course, yet more complex forms of coordination may be required to perform effectively on joint tasks.

Second, our results indicate that behavioral coordination does not invariably play a role in generating positive affect. To result in liking, coordination needs to involve those behaviors that are relevant to people's interaction goals and concerns. In Experiment 1, coordination in movement was critical to task success and thus promoted liking for the virtual partner, whereas coordination in color changes was irrelevant to success at the task and did not have an impact on liking. In both experiments, liking went beyond an expression of affect to include inferences regarding the likely effectiveness of the partner in performing a subsequent task. Coordination-induced liking, in other words, is valued because it is perceived to be a foundation for effective joint action with another person.

In view of the adaptive value of interaction fluency, one might ask why the experience of coordination is not ubiquitous in social life. Why don't people invariably find ways to integrate their actions over time, and, as a consequence, feel positive toward one another? There are at least two sets of reasons why this sanguine state of affairs is not always (or even commonly) obtained. First, coordination assumes compatibility of goals and of actions to achieve these goals. Coordination is likely to be facilitated by cooperative orientation, for example, but is likely to be hindered by conflict (cf. Deutsch 1973). Clearly, if we perceive our goals to be in conflict with those of someone else, coordination in pursuit of mutual goal attainment is unlikely.

Second, coordination can take complex forms that are difficult to achieve and maintain, even when the individuals involved have a cooperative orientation. Everyday life is replete with contexts involving difficult or complex activities (e.g., work groups, sports teams, ad hoc committees) in which people attempt to coordinate their efforts but fail to do so. Coordination in such instances may be temporarily enhanced by external cues and forces, but in the absence of such factors the individuals may fail to maintain their interaction fluency. Even with strong influence, moreover, complex forms of coordination may be difficult to achieve in dyads and groups. An interesting issue for future research is the relationship between coordination complexity and positive affect. Perhaps the more complex the form of coordination the effect on liking is the stronger.

We have developed a formal model, instantiated in computer simulations, that identifies key factors responsible for the achievement of simple and complex forms of coordination (Nowak et al. 2002; Vallacher et al. 2005). In this model, each individual follows to a certain degree their own internal dynamics and to a certain degree reacts to the state of the other individual with whom they are interacting. The strength of influence from the partner is systematically varied, as is the degree of similarity in the internal dynamics of the two individuals. Computer simulations revealed that for partners with vastly different internal dynamics, coordination could be achieved only under conditions of strong mutual influence, and only relatively simple forms of coordination (e.g., in-phase synchronization) were observed. When the internal dynamics of the partners were similar, however, minimal influence between the individuals was sufficient for the development of behavior coordination, and the coordination achieved could take very complex forms and could be sustained over long periods of time. We also found that when systems were initially coordinated by strong influence, they failed to develop similar internal parameters. As a result, when the influence was removed, the systems immediately became decoupled in their dynamics and displayed little or no subsequent behavior coordination.

Taken together, the simulation results suggest that when the required coordination is complex (e.g., involving anti-phase relations or a set of distinct actions) rather than simple (e.g., mimicry), it is necessary for the individuals to have (or develop) compatible internal dynamic tendencies. Strong influence may promote coordination, but there is a limit to the complexity of coordination that can be achieved in this manner, and this achievement is unlikely to be sustained when the influence is withdrawn or otherwise dissipated. In essence, effective and enduring forms of behavioral coordination occur under relatively weak influence, with individuals developing internal parameters (e.g., emotions, values, interpretative sets) that enable them to display spontaneous interpersonal fluency by virtue of having a common wellspring for action.

The present research represents an initial step in validating the hypothesized signaling and also the functional role of coordination in social relations. Future research is necessary to establish the generality of the effects observed in the experiments and to test other implications of the theoretical rationale. We suggest, for example, that the association between interaction fluency and liking is mediated by the functional value of coordination for dyadic or group goal attainment. Several implications amenable to empirical test follow from this assumption. Consider, for example, two people who achieve spontaneous coordination in their behavior while performing a task, but the task ends in failure. If coordination and liking were linked by a self-contained positive feedback loop, the failure might have little consequence – indeed, it might strengthen the relationship by emphasizing the intrinsic as opposed to the outcome-oriented nature of the relationship (cf. Seligman et al. 1980). However, if the affect-generating value of interaction fluency is based on an implicit link between coordination and adaptive functioning, then clear indications of failure might promote strong negative affect between two people who are well coordinated. The recognition that one is coupled with someone who is likely to promote ineffective collective action should launch efforts to terminate the relationship. Failure on the part of weakly coordinated individuals, in contrast, might inspire efforts to increase the quality of their coordination, possibly increasing their liking for one another.

We are currently investigating these and other implications of our rationale concerning social coordination and relationship formation. The underlying premise of this work is that social relations reflect principles of dynamical systems that are invariant across psychological phenomena (cf. Nowak and Vallacher 1998; Vallacher and Nowak 2007). Thus, perceptual, cognitive, and action systems attain fluency as their component elements achieve coordination, often by means of self-organization, in the pursuit of a higher-order state. In a similar fashion, social relations attain interpersonal fluency as the component elements (i.e., individuals) achieve coordination in their respective internal states and behavior patterns. By framing social dynamics in this fashion, one can envision a science of social psychology that is internally coordinated and reflects theoretical fluency.

References

Abele, S., Stasser, G.: Coordination success and interpersonal perceptions: matching versus mismatching. J. Pers. Soc. Psychol. **95**, 576–592 (2008)

Ashton-James, C., van Baaren, R. B., Chartrand, T. L., Decety, J., Karremans, J.: Mimicry and me: The impact of mimicry on self-construal. Social Cognition, **25**, 518–535 (2007)

Bargh, J.A., Chen, M., Burrows, L.: Automaticity of social behavior: direct effects of trait construct and stereotype activation on action. J. Pers. Soc. Psychol. **71**, 230–244 (1996)

Baron, R.M., Amazeen, P.G., Beek, P.J.: Local and global dynamics of social relations. In: Vallacher, R.R., Nowak, A. (eds.) Dynamical Systems in Social Psychology, pp. 111–138. Academic, San Diego (1994)

Bernieri, F.J., Reznick, J.S., Rosenthal, R.: Synchrony, pseudosynchrony, and dissynchrony: measuring the entrainment process in mother–infant interactions. J. Pers. Soc. Psychol. **54**, 243–253 (1988)

Chartrand, T.L., Bargh, J.A.: The chameleon effect: the perception–behavior link and social interaction. J. Pers. Soc. Psychol. **76**, 893–910 (1999)

Condon, W.S.: Cultural microrhythms. In: Davis, M. (ed.) Interaction Rhythms: Periodicity in Communicative Behavior, pp. 53–76. Human Sciences Press, New York (1982)

Condon, W.S., Ogston, W.D.: Sound film analysis of normal and pathological behavior patterns. J. Nerv. Ment. Dis. **143**, 338–347 (1966)

Coovert, M.D., Reeder, G.D.: Negativity effect in impression formation. The role of unit formation and schematic expectation. J. Pers. Soc. Psychol. **26**, 49–62 (1990)

Csikzentmihalyi, M.: Flow: The Psychology of Optimal Experience. Harper & Row, New York (1990)

Decéty, J., Chaminade, T.: Neural correlates of feeling sympathy. Neuropsychologia: Spec. Issue Soc. Cognit. **41**, 127–138 (2003)

Deutsch, M.: The Resolution of Conflict: Constructive and Destructive Processes. Yale University Press, New Haven (1973)

Dijksterhuis, A., Bargh, J.A.: The perception–behavior expressway: automatic effects of social perception on social behavior. In: Zanna, M.P. (ed.) Advances in Experimental Social Psychology, vol. 33, pp. 1–40. Academic, San Diego, CA (2001)

Harakeh, Z., Engels, R.C.M.E., van Baaren, R.B., Scholte, R.H.J.: Imitation of cigarette smoking: an experimental study on smoking in a naturalistic setting. Drug Alcohol Depend. **86**, 199–206 (2007)

Hatfield, E., Cacioppo, J. T., Rapson, R. L.: Emotional contagion. Cambridge University Press, New York (1994)

Isen, A. M.: The influence of positive and negative affect on cognitive organization: Some implications for development. In N. Stein, B. Leventhal, T. Trabasso (eds.), Psychological and biological approaches to emotion, pp. 75–94. Erlbaum, Hillsdale, NJ. (1990)

Jirsa, V.K., Kelso, J.A.S. (eds.): Coordination Dynamics: Issues and Trends. Springer-Verlag, Berlin/Heidelberg (2004)

Kaneko, K. (ed.): Theory and Applications of Coupled Map Lattices. World Scientific, Singapore (1993)

Kanouse, D.E., Hanson, L.R.: Negativity in Evaluations. General Learning Press, Morristown (1971)

Kawakami, K., Young, H., Dovidio, J.F.: Automatic stereotyping: category, trait, and behavioral activations. Pers. Soc. Psychol. Bull. **28**, 3–15 (2002)

Kelso, J.A.S.: Dynamic Patterns: The Self Organization of Brain and Behavior. MIT Press, Cambridge, MA (1995)

Kendon, A.: Movement coordination in social interaction: some examples. Acta Psychol. **32**, 1–25 (1970)

Kulesza, W., Nowak, A.: Lubię Cię, bo jesteśmy dobrze zgrani: wpływ koordynacji na pozytywne nastawienie w relacjach społecznych. Przegląd Psychologiczny **46**, 323–338 (2003) [I like you because we are tuned together well: influence of coordination on positive attitude in social relations. Psychol. Rev.]

LaFrance, M.: Nonverbal synchrony and rapport: analysis by the cross-lag panel technique. Soc. Psychol. Q. **42**, 66–70 (1979)

Lankin, J.L., Jefferis, V.E., Cheng, C.M., Chartrand, T.L.: The chameleon effect as social glue: evidence for the evolutionary significance of nonconscious mimicry. J. Nonverbal Behav. **27**, 145–162 (2003)

Lewenstein, M., Nowak, A.: Recognition with self-control in neural networks. Phys. Rev. A **40**, 4652–4664 (1989)

Marsh, K.L., Richardson, M.J., Baron, R.M., Schmidt, R.C.: Contrasting approaches to perceiving and acting with others. Ecol. Psychol. **18**, 1–38 (2006)

Meltzoff, A.N., Moore, M.K.: Imitation of facial and manual gestures by human neonates. Science **198**, 75–78 (1977)

Newtson, D.: The perception and coupling of behavior waves. In R. R. Vallacher, A. Nowak (eds.). Dynamical systems in social psychology, pp. 139–167. : Academic Press, San Diego, CA (1994)

Nowak, A., Vallacher, R.R.: Dynamical Social Psychology. Guilford Press, New York (1998)

Nowak, A., Vallacher, R.R.: The emergence of personality: dynamic foundations of individual variation. Dev. Rev. **25**, 351–385 (2005)

Nowak, A., Vallacher, R.R., Zochowski, M.: The emergence of personality: personal stability through interpersonal synchronization. In: Cervone, D., Mischel, W. (eds.) Advances in Personality Science, vol. 1, pp. 292–331. Guilford Publications, New York (2002)

Nowak, A., Vallacher, R.R.: Synchronization dynamics in close relationships: coupled logistic maps as a model for interpersonal phenomena. In: Klonowski, W. (Series ed.) Frontiers on Nonlinear Dynamics. From Quanta to Societies, vol. 2, pp. 165–180. Pabst Science Publishers, Berlin (2003)

Pratto, F., John, O.P.: Automatic vigilance: the attention grabbing power of negative information. J. Pers. Soc. Psychol. **61**, 380–391 (1991)

Schmidt, R.C., Carello, C., Turvey, M.T.: Phase transitions and critical fluctuations in the visual coordination of rhythmic movements between people. J. Exp. Psychol. Hum. Percept. Perform. **16**, 227–47 (1990)

Seligman, C., Fazio, R.H., Zanna, M.P.: Effects of salience of extrinsic rewards on liking and loving. J. Pers. Soc. Psychol. **38**, 453–460 (1980)

Shockley, K., Santana, M.V., Fowler, C.A.: Mutual interpersonal postural constraints are involved in cooperative conversation. J. Exp. Psychol. Hum. Percept. Perform. **29**, 326–332 (2003)

Simon, H.A.: Motivation and emotional controls of cognition. Psychol. Rev. **74**, 29–39 (1967)

Simpson, J.A., Orina, M.M., Ickes, W.: When accuracy hurts, and when it helps: a test of the empathic accuracy model in marital interactions. J. Pers. Soc. Psychol. **85**, 881–893 (2003)

Strack, F., Martin, L.L., Stepper, S.: Inhibiting and facilitating conditions of the human smile: a nonobtrusive test of the facial feedback hypothesis. J. Pers. Soc. Psychol. **54**, 768–777 (1988)

Strogatz, S.H.: Sync: The Emerging Science in Spontaneous Order. Hyperion, New York (2003)

Tanner, R.B., Ferraro, R., Chartrand, T.L., Bettman, J.R., van Baaren, R.B.: Of chameleon and consumption: the impact of mimicry on choice and preferences. J. Consum Res **34**, 754–766 (2007)

Tesser, A.: Self-generated attitude change. In L. Berkowitz (ed.) Advances in Experimental Social Psychology, Vol. 11, pp. 289–338. Academic Press, New York (1978)

Turvey, M.T.: Coordination. Am. Psychol. **4**, 938–953 (1990)

Vallacher, R.R.: Mental calibration: forging a working relationship between mind and action. In: Wegner, D.M., Pennebaker, J.W. (eds.) The Handbook of Mental Control, pp. 443–472. Prentice-Hall, New York (1993)

Vallacher, R.R., Nowak, A.: Dynamical social psychology: finding order in the flow of human experience. In: Kruglanski, A.W., Higgins, E.T. (eds.) Social Psychology: Handbook of Basic Principles, 2nd edn, pp. 734–758. Guilford Publications, New York (2007)

Vallacher, R.R., Wegner, D.M.: A Theory of Action Identification. Erlbaum, Hillsdale (1987)

Vallacher, R. R., Read, S. J., Nowak, A.: The dynamical perspective in personality and social psychology. Pers. Social Psychol. Rev., 6, 264–273 (2002)

Vallacher, R.R., Nowak, A., Kaufman, J.: Intrinsic dynamics of social judgment. J. Pers. Soc. Psychol. **67**(1), 20–34 (1994)

Vallacher, R.R., Nowak, A., Zochowski, M.: Dynamics of social coordination: the synchronization of internal states in close relationships. Interact. Stud. **6**, 35–52 (2005)

van Baaren, R.B., Maddux, W.W., Chartrand, T.L., de Bouter, C., van Knippenberg, A.: It takes two to mimic: behavioral consequences of self-construals. J. Pers. Soc. Psychol. **84**, 1093–1102 (2003)

van Baaren, R.B., Holland, R.W., Kawakami, K., Knippenberg, A.V.: Mimicry and prosocial behavior. Psychol. Sci. **15**, 71–74 (2004) van Leeuwen, M.L., Velig, H., van Baaren, R.B., Dijskterhuis, A.: The influence of facial attractiveness on imitation. J. Exp. Soc. Psychol. **45**, 1295–1298 (2009)

Wilson, M., Knoblich, G.: The case for motor involvement in perceiving conspecifics. Psychol. Bull. **131**, 460–473 (2005)

Wiltermuth, S. S., Heath, C.: Synchrony and cooperation. Psychol. Science, 20, 1–5 (2009)

Winkielman, P., Cacioppo, J.T.: Mind at ease puts a smile on the face: psychophysiological evidence that processing facilitation increases positive affect. J. Pers. Soc. Psychol. **81**, 989–1000 (2001)

Winkielman, P., Schwarz, N., Nowak, A.: Affect and processing dynamics. In: Moore, S., Oaksford, M. (eds.) Emotional Cognition, pp. 111–138. Benjamins, Amsterdam (2002)

Wojciszke, B.: Lubienie, respekt i percepcja polityczna. Studia Psychologiczne **41**, 49–74 (2003). Liking, respect and political perception. Psychol. Stud

Zajonc, R.B.: Attitudinal effects of mere exposure. J. Pers. Soc. Psychol. **9**, 1–27 (1968)

Zajonc, R.B.: Feeling and thinking. Am. Psychol. **35**, 151–175 (1980)

Zochowski, M., Liebovitch, L.: Synchronization of trajectory as a way to control the dynamics of the coupled system. Phys. Rev. E. **56**, 3701 (1997)

Chapter 11
The Dynamics of Trust from the Perspective of a Trust Game

Magda Roszczynska-Kurasinska and Marta Kacprzyk

Abstract Trust is an essential feature in the majority of everyday social activities. Trust is not static but dynamic; it changes over time, manifesting itself differently in different situations and with different behaviors of others. Many of these changes can be captured in experimental settings based on a 'trust game,' coming from game theory. This game enables us to explore the consequences of various experimental manipulations of the many different variables that affect the degree of trust. However the mechanisms captured by the trust game have not yet been described from a dynamic systems perspective. In this chapter this deficiency is rectified.

Introduction: From Risk to Trust

Uncertainty is an inherent, unavoidable element of human life. As the philosopher Pliny the Elder argued "The only certainty is that nothing is certain"[1] and in our everyday experience this statement is sometimes painfully true. Not knowing what will happen next, we are kept in suspense. In such challenging situations people frequently experience fear. They attribute their tension to, e.g., the global changes, job/health/love/weather or insecurity, and tend to ask "what if ... were to happen," worrying about the answer. However, while anxiety is adaptive in that it focuses attention, living in constant anxiety is not that useful, so the human species,

---

[1] Albert Einstein was somehow more successful in finding inevitable things in our reality claiming that "two things are certain: the universe and human stupidity" although he was not certain about the universe.

M. Roszczynska-Kurasinska (✉)
Institute for Social Studies, University of Warsaw, Stawki 5/7, Warsaw 00183, Poland
e-mail: magda.roszczynska@gmail.com

M. Kacprzyk
Institute for Social Studies, University of Warsaw, Stawki 5/7, Warsaw 00183, Poland
e-mail: marta.kacprzyk@gmail.com

A. Nowak et al. (eds.), *Complex Human Dynamics*, Understanding Complex Systems, 191
DOI 10.1007/978-3-642-31436-0_11, © Springer-Verlag Berlin Heidelberg 2013

in the course of its evolution, developed various ways of coping with major uncertainty, specially with uncertainty related to the behavior of other people.

Acceptance (to some extend) of uncertainty, the unpredictability of events, risk and change is a prerequisite but an insufficient condition for dealing with reality, managing our lives as well as achieving our goals in society. A basic tolerance of uncertainty may be supplemented by various strategies people have learned to exploit. One commonly used is the search for relevant information which will help in making better decisions. However, as in complex problems we usually face limits both of time and of our cognitive resources, it is impossible to collect all the information necessary.

Uncertainty can also be reduced through extensive (and usually expensive) control mechanisms – probably most popular among advocates of totalitarianism. Nevertheless, historically total control over man has proved hard (if not impossible) to impose and new revolutions and rebels prove totalitarianism shortcomings and ineffectiveness.[2]

Both these strategies require the considerable utilization of different kinds of resources (e.g., money, time) at the individual and collective level. However, there is another way of coping with uncertainty that helps to save these resources – trust, a useful although sometimes not 100 % reliable means. By removing the need for checking-up and the laborious verification of other people, trust enhances cooperation. We encounter trust in everyday situations, e.g. when we cross a road intersection on the green light, when we follow the medical advice of a physician, when we act upon the recommendation of a friend. As Luhmann (1968) pointed out: "Man has admittedly in many situations a choice whether or not to put his trust forward in a certain way. Without any trust, however, he could not leave his bed in the morning." But what exactly is trust?

There is an abundance (maybe even an overabundance) of definitions, as well as large body of literature illustrating different forms of trust. There is no common definition that would be accepted by all researchers coming from diverse scientific backgrounds. The lack of universal definition of trust may lead to confusion between psychologists, sociologists, economists, etc. It might be wise to be precise and use different terms for different notions of trust, distinguishing trust from 'reliance,' 'confidence', 'faith', 'hope' and 'familiarity' (McLeod 2011; Baier 1986; Sztompka 2007; Luhmann 1968. There are some such attempts but still they lack coherence, as there is no agreement between scientists on their different meanings. For the purpose of this chapter we have chosen a simple definition of trust: "Trust is the expectation that a partner will not engage in opportunistic behavior, even in the face of opportunities and incentives for opportunism" (Bachmann and Zaheer 2006, p. 252). We feel that this restricted definition meets our understanding of 'pure trust' that most people share.

---

[2] In hierarchical organizations a similar notion of control is used. Here yet another interesting mental process takes place, not at the top where the power is concentrated, but at lower levels where the power is executed. It manifests itself as a comforting thought that someone else makes the decisions and takes the responsibility for us. Shifting responsibility is a common practice that hinders the growth of many organizations.

This chapter aims at exploring the different manifestations and conditions for expectations of a trust-giver (trustor) and a trust-taker (trustee) from a dynamic perspective. It deals mainly with interpersonal trust, the central paradigm of trust research (McLeod 2011). The chapter is structured as follows. We begin by introducing the concept of trust as a dynamic phenomena, chosen as it captures the complex nature of trust. We then present the different methods used for measuring trust, comparing static and dynamic approaches. One of these methods is particular interesting and we devote the rest of the chapter to it. This method is a trust game which permits testing diverse trust situations by implementing just small changes in settings of the game, as will be seen in our discussion of the results of various trust game experiments which reveal manifold trust factors. We analyze these outcomes showing how they can be interpreted and adjusted in dynamic perspective. We conclude by showing the limitations of the trust game.

Trust as a Dynamic Phenomenon

Trust is dynamic. Trust changes with new experiences, with transformations of the environment, with the emotional state of individuals etc. Our willingness to place trust in others evolves in time: appears, develops (grows or shrinks), sometimes expires. New interaction, new information may affect our trust decisions. This temporal dimension means that trust is a variable and complex phenomenon, sometimes too elusive to capture its entire character.

We can describe trust in a more formal way, as a dynamical system. In the phase space in which trust operates, there will be regularities due to attractors. Everyone has their level of trust which shapes their expectations towards others – some people trust more, some less. Some trust attractors are very strong, while others may be weak (Fig. 11.1). For example, many researchers (as well as everyday experience) show that distrust is a strong attractor from which it is hard to get out. Intractable conflicts illustrate this (see Bui-Wrzosinska, this volume, Chap. 13).

An individual's disposition towards trust is usually stable over time. This does not mean that people with a high level of trust place trust in everybody, no matter what. We usually have a set of attractors that are 'activated' in different situations, e.g., one can fully trust the physicians that their medical diagnoses are correct, but at the same time distrust the government for their economic prognoses and not follow our neighbor's gardening tips.

Factors Shaping Trust

The attractors for trust may be described by two main characteristics: bi-modality and strength. Bi-modality refers to a person's expectations (and therefore decisions) on whether to trust or not; strength determines the stability of the attractor.

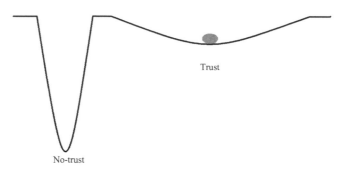

Fig. 11.1 The figure depicts exemplary attractors of 'trust' and 'no-trust'. The state of a person in a given moment is indicated by the *dot*. For a time the person is in the 'trust' attractor. This attractor is broad but shallow, meaning that in most situations the person would prefer to trust; however, this state is very sensitive to external and internal factors. It is relatively easy to push the person to the 'no-trust' attractor which is very deep and it would be relatively hard to push the person out of it

Both characteristic are influenced by various factors such as genes, gender, emotions, experience, individual heuristics, environment etc. These factors are the subject of interest of much research on trust. The following examples, as well as many others, shape our trust behaviors:

- Genes: "Evolutionary psychology suggests that a tendency towards 'give and take' (reciprocity), and accepting relational risk, is 'in our genes', since it was conducive to survival in the ancient hunter – gatherer societies in which humanity evolved." (Bachmann and Zaheer 2006, p. 736)
- Emotion: "Happiness and gratitude, emotions with positive valence, increase trust, and anger, an emotion with negative valence, decreases trust" (Dunn and Schweitzer 2005, p. 253). However, Dunn and Schweitzer's analysis shows that emotions do not play a significant role when it comes to people familiar to the trustor and when the trustor is aware of the source of those emotions.
- Experience also shapes peoples predisposition to trust. According to dynamical approach "small, random events early in the history of a system determine the ultimate end state, even when all end states are equally likely at the beginning" (Sterman 2000, p. 349).
- Social, political and economical climate also contributes to how people trust one another (McLeod 2011). For example it is believed that democracy is the political system in which people can trust others more often than in totalitarian ones (Uslaner 1999) as the oppressive systems make it rather irrational to trust those who oppress (Baier 1986; Potter 2002).

How to Measure Trust

Just as there are many factors that affect our decision whether to place trust in somebody there are also many ways of measuring individual trust. Generally we can distinguish two main approaches to measuring trust: static vs. dynamic. Each of the approaches contributes to the theoretical and practical understanding of trust and its role in social relations. To illustrate the difference between them in a simple way, assume that the static approach advocates the use of manifold types of tools depicting the state in a given moment. This approach enables the elicitation of the structural characteristics of trust. For example, interpersonal trust measures created by McAllister (1995) can be used to investigate the relationship between cultural-ethnic similarity and cognition-based trust using regression analysis. On the other hand, the dynamical approach investigates the trust and factors shaping it using tools that capture changes in time. We can, for instance, conduct longitudinal studies or design quasi-experiments where time and change in time are considered.

In the design of many experiments conducted in the dynamical perspective, a game theoretical approach is used. In pure game theory the motivation of players is reduced to maximization of profits, setting aside other features that shape human behavior. The actions chosen by players are determined by the expected output. It is assumed that players will cheat when cheating will be more rewarding or honor trust when, in given context, this strategy will give them better outcomes. This is 'calculative trust' where trust is a mode of behavior that is consciously chosen after a rational assessment of the profits and costs of trust (James 2002).

Axelrod's work on the Prisoner's dilemma, in which the dynamic nature of cooperation was perused (Axelrod 1984), have inspired many trust researchers. In general, the Prisoner's dilemma is a formal description of situation in which an individual's rational strategy is to cheat even though a cooperative strategy by both players would result in superior payoffs. Moreover, a cooperative strategy is more fragile since each player's incentive is to behave opportunistically and exploit the other if possible. The settings of this dilemma are simple: if one person plays cooperatively, then the other can defect and collect the biggest possible payoff, leaving the 'naive' other with the lowest possible payoff. Expecting this course of events both sides usually decide to defect which leads them both to have the sub-optimal results. The collectively best outcome, achievable only when both sides cooperate, is lost, depriving both sides of its benefits. The Prisoner's dilemma shows that a strategy that avoids risk for the individual is not always effective, certainly not for the group, but also not for the individuals.

Another interesting example of a game that enables us to explore the trust behavior in the theoretical framework of game theory is the 'trust game.' This game has significantly deepened our understanding of the role of reciprocity, expectation and experience in the development of trust. We devote the following sections to this game.

Introduction to the Trust Game

In recent times, researchers interested in measuring individual willingness to trust, as well as trust reciprocation, have been using a new tool, the 'trust game.' With the introduction of sequentiality, the trust game has become one of the standards for measuring trust and is broadly used to examine the emergence of cooperative behaviors and motivations behind them. The advantage of this method is its simplicity and relative ease by which various outcomes can be explored.

In the classic version of the trust game an individual (called the trustor) chooses whether to place trust in second player by deciding how much money, if any, from an initial endowment to send to an anonymous partner (in some experimental settings the amounts of money that can be transferred are fixed). This amount is then augmented (usually tripled) by the experimenter and passed to the partner. The second player (trustee) who received the multiplied sum can then decide whether to reciprocate the trust or defect. In some experimental settings of the trust game individuals take on roles of trustor and trustee in an alternating sequence till one of them refuses to return the money; in some the number of stages is limited, in others, it is not.

The classic trust game may have the following structure: both players receive, say, $10 ($10, $10). Let's assume that the initial decision maker passes the entire sum to the partner. The second player receives $30 and faces the dilemma of whether to keep all the money for themselves or not. If they keep they end up with $40 (0, $40) – how not to succumb to temptation like this! What is the best strategy for the trust game? According to game theory, the Nash equilibrium for purely selfish preferences is to send nothing as a selfish partner would probably choose not to reciprocate our gift. Realizing this, the first player refuses to engage and the equilibrium payoff is settled ($10, $10). Contrary to this, in research applications, some people do engage in the game and honor trust. When the trustor sends some money to the trustee, the most popular behaviors of trustee, apart from not sending anything back, are: to reciprocate and pass the half of the received amount (in our example the payoff would be ($15, $25)) or even 2/3 of the received sum ($20, $20).

This counter-rational behavior leads us directly to the question: Why is the trust game so appealing? Exploration of motivations behind the non-selfish decisions is the most absorbing part of the trust game, shedding light on the nature and dynamics of trust. Furthermore the simplicity of the structure permits the investigation of the relations of the results on the level of payoffs. By changing the settings, one can easily explore the modifications of behavior that result if we, e.g., increase or decrease the payoffs, introduce 'social history' (Berg et al. 1995) or change the social distance between the players.

The trust game scheme is not the only means to analyze trust, but it reflects our every day experience, in particular circumstances where a two-person exchange takes place. It represents many situations in which "the attractiveness to one party of a welfare-increasing investment hinges on the trustworthiness of another" (Bachmann and Zaheer 2006, p. 56). It is easy to think of an example, such as a person who purchases goods via the internet from unknown seller trusting that they will indeed receive the goods in a later shipment; or a child who helps to clean the house trusting that they will be allowed to go to the movies later. In McEvily et al. (2006) we find a particularly appealing example where the owner of a small company has a dilemma: how much money to invest in the training of her

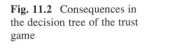

Fig. 11.2 Consequences in the decision tree of the trust game

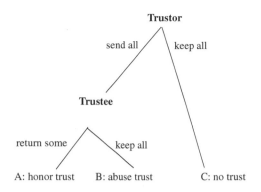

employees, reasoning that the more qualified workforce the better might be the profits. However, after completing an expensive training an employee may consider whether the prospects for their future in the current company are promising enough or whether it would be more beneficial to go to another firm. If the second option is more attractive the well-trained employee is leaving. This situation is close (if not the same) as presented in the trust game. All of these diverse situations can be modeled by one tool, the two-person trust game.

What does the Trust Game Measure?

In this subsection we will look closer at the formal presentation of the trust game. The decision tree in Fig. 11.2 depicts all possible actions of both players and the outcomes of these actions. The value of the outcome is not fixed; it can be changed according to the hypotheses being tested. The exemplary payoffs depending of the actions taken by the players is illustrated in Table 11.1. Changing the payoffs results in the calculation of complex factors that influence the trustors' and trustees' decisions in trust game. These factors should be taken into account, especially in a situation when an analyst is comparing the results from two or more experiments with different payoffs. The most frequently investigated variables include: incentive, guilt, temptation and risk. Below we present a brief description of them and the methods for their calculation.

Incentive to Abuse Trust

The incentive to abuse trust is a variable that negatively influences the willingness to reciprocate the trust by the trustee. It is measured as a difference between the payoff of the trustee in a situation when he abuses trust (B2) and his payoff when he reciprocates it (A2). The larger the difference between B2 and A2 the greater is the incentive to abuse trust and higher the chances that the trustee will behave opportunistically.

Table 11.1 Exemplary payoffs of players initially given $10 each, depending on behavior

Behavior	Player	Payoff
Trust reciprocated	Trustor	A1 = $20
	Trustee	A2 = $20
Trust abused	Trustor	B1 = $0
	Trustee	B2 = $40
No trust	Trustor	C1 = $10
	Trustee	C2 = 10

Guilt

The guilt is experienced by a trustee who does not want to respect the trust, but paradoxically, anticipation of guilt positively influences their willingness to reciprocate the trust. It is measured as a difference between the expected payoffs to the trustor (B1) and the trustee (B2) in a situation when the trustee abuses trust. When this difference is small the feeling of guilt is not so high and the trustee can be tempted to behave untrustworthily. However, if their payoff is much higher than trustor's, the guilt could prevent them from abusing trust. The guilt decreases the utility of the payoff gained from untrustworthiness and can result in actions honoring trust.

Risk

In a trust game the risk is experienced only by the trustor in a situation of placing trust. The value of risk can be assessed in two ways. First, it can be measured as the amount of money sent to the trustee, sometimes called 'opportunity cost.' Although in the game setting depicted in Fig. 11.2 the trustor has only two options: to send all the money to the trustee or keep it all, in other versions of the game the trustor can choose the proportion of money they want to sent to the trustee and therefore declare how much risk they are ready to take. Second, risk can be measured as a proportion between two differences: the first difference is that between how much the trustor will retain when they do not place trust in trustee compared to when they place trust but it is abused (C1 − B1); the second difference is that between how much they can gain once the trust is honored by the trustee and how much they retain when it is abused (A1 − B1). The lower the proportion (C1 − B1)/(A1 − B1) the lower the risk. The numerator can be understood as the potential maximum loss for the trustor. The worst outcome for trustor when he does not place trust is C1, and the worst outcome of the game when he places trust is B1, the situation when their trust is abused The higher the potential maximum loss the higher is the risk. The dominator, on the other hand, can be interpreted as the potential maximum gain from the game once the trustor places trust. The higher it is the lower is the risk.

Temptation

Temptation is assessed as the ratio of the incentive to abuse trust (B2 − A2) to the guilt resulting from abusing trust (B2 − B1). The higher is the incentive to abuse trust and lower the guilt from doing so, the higher is temptation. A high incentive to behave untrustworthily can be off-set by a significant level of the guilt which would result from a big disproportion between the trustee's and trustor's payoffs.

Other factors influencing trust decisions in a trust game have been used, but the ones presented above are the most frequent.

What Can We Learn from the Trust Game Experiments?

In this chapter we present experiments which demonstrate how useful the trust game is for examining trust. By using different settings of the trust game, it is easy to test almost any hypothesis referring to the evolution of trust in a two-person transaction environment and investigate the influence of different factors on the process of decision making involving trust and reciprocity. Thanks to the trust game it is also possible to capture the dynamics of trust. However, our thorough analysis of the data obtained in research using the trust game shows that they are seldom interpreted from the perspective of dynamical systems. Here, we will present not only examples of experiments in which the trust game was used but also provide interpretation of the results showing the possible attractors of trust/reciprocity and the mechanisms that shape them. Moreover, we will present a way of adjusting the experimental settings so that the results are more suitable for interpretation in a dynamical approach.

Before providing a detailed description and analysis of the findings of these experiments as dynamical systems, we present the general outcome of the game when the players do not know anything about each other. In this pure setting trustors place trust in trustees in 37 % of the trust games, no matter the amount of the monetary payoff (Snijders 1996). This trust is reciprocated by trustees in 36 % of these trust games in which trust was granted. We can interpret this data saying that for majority of the people the dominant attractors are for the trustor to place no trust and for the trustee to abuse trust, whereas to place trust and to honor it are attractors which occur less often.

The attractors and the phase space to which they belong can, however, be easily changed once specific information about various relationships between players is disclosed. First, we will show how the identification of the partner's intentions enhances the willingness of reciprocity. Second, we introduce data indicating the importance of past experience on trust and trustworthiness. Third, we present the research testing the influence of playing both roles in the trust game on decisions whether or not to place trust and also on whether or not to reciprocate trust.

How Intentions of the Partner Influence Willingness for Reciprocity

Society is structured along reciprocal rules. Principles such as "an eye for an eye", "a favor for a favor", "give and take", "you scratch my back and I'll scratch yours" have been known all around the world for thousands of years. Usually there are no written rules regulating when and how to reciprocate, yet we use decision heuristics that help us in such situations intuitively. However, reciprocity is not a good strategy in every situation; application of this strategy depends on specific characteristic of the interaction.

To what extend can one expect that one's, say, positive behavior toward a stranger will be reciprocated? This was exactly the question asked by McCabe et al. (2003). They assumed that the reciprocity of trusting behavior depends on the analysis of alternatives that the trustor has. They wanted to check if a reduction of trustor's opportunity cost in the trust game influences the reciprocity behavior of trustee. It is believed that humans interpret the behaviors of others, attributing to them specific beliefs, desires, emotions and so on, and behave accordingly. On the assumption that, when somebody voluntarily places trust in us, it is socially obligator to reciprocate, they thought that attribution of a free will to the trustor would change the level of reciprocity of a trustee in the trust game. However, where the trustor sends the trustee money because they have no option, the trustee will behave selfishly since the trustor did not bear any opportunity cost and, according to the trustee, they do not need to be compensated for this compulsory behavior.

To test this hypothesis they used a modified version of the trust game. By informing the trustee that the trustor had to pass money (they could not keep all the money for themselves), they restricted the trustee's ability to read the trustor's intentions unambiguously, while in the classical setting of the trust game the intentional move of sending money can be interpreted by the trustee as an act of trust.

Analysis of the data shows that, while making the decision on whether or not to reciprocate, the decoding of the intentions of the counterparty plays an important role. It is not the pure economic outcome that mostly influences the decisions of reciprocating the action but the ability of reading the motives of the trustor. People reciprocate the trust more often in the situation when they know that the trustor voluntarily placed trust in them.

These results shed light on the mechanism of reciprocity. McCabe et al. showed that the way we interpret the reality and behaviors of others makes an important change to the way we act towards others who previously placed trust in us, yet they did not contribute to explaining the dynamics of this change. To provide the analysis missing from the perspective of dynamical systems, one should reconstruct the experimental design to introduce one more variable, time. The simplest change of the procedure would allow subjects to play the trust game more than once, say ten times. This modification would result in measuring the acts of reciprocity in time and detect individual tendencies for reciprocity depending on the setting.

We assume that everybody has attractors for reciprocity and no-reciprocity. The shape of these attractors depends on individual predispositions; both attractors do not necessarily have the same depth. People fall into one of two attractors, depending on the situation. The attractor the person falls into in a given moment is shaped by many factors. One of them may be the interpretation of behaviors of others. We would hypothesize, based on the results of McCabe et al., that a person's interpretation of trust behavior as a voluntary action is a factor strong enough (but not for all subjects) to push this person out from the attractor of 'abusing trust' and push him/her into the attractor of 'honoring trust'. Of course if the person was in the attractor of 'honoring trust' before making the interpretation nothing would change in terms of observable behavior, although a latent change could happen. This change would result in the reshaping of the current attractor which could be deepened and stronger forces would be needed to push the person out of it.

Beside the interpretation of intentions, there are many other factors that influence the phase space of trust and reciprocity. The one that imprints itself significantly on the properties of the attractors is an experience to which we devote the next subsection.

Experience as a Factor Reducing or Enhancing Trust and Trustworthiness

Imagine a situation in which you need to trust a stranger. Probably, before making up your mind whether to trust or not, you ask yourself many questions such as: does the stranger look trustworthy, how much can I lose if my trust would be disregarded, what other choices do I have, and so on. Although finding answers to these questions in most situations is rather easy, understanding the mechanism underlying them is much more complicated. Let's take, for instance, the evaluation of potential loss. The value of loss is not an objective value but rather a subjective one which can depend on subject's general wealth, the context of the situation, etc. (Khaneman and Tversky 1979).

Here, we will concentrate on trying to understand the factors influencing the estimation of a stranger's trustworthiness, especially the influence of past experience. We will present two experiments aimed at understanding the effect of past experience on decisions involving trust: (1) an experiment by Rigdon et al. (2007) and (2) an experiment by McEvily et al. (2006) (we call them Experiment 1 and Experiment 2 respectively). In both experiments subjects were asked to play the trust game more than once with an anonymous counterpart. In the first experiment subjects did not know anything about their counterparts, while in the second one subjects were assigned randomly to one of two groups (group A, group B) at the beginning of the game.

Description of Experiment 1

Rigdon et al. (2007) hypothesized that people would exhibit more trust and trust-worthiness in an environment in which they themselves experience more trust/trustworthiness compared to their standard level of trust/trustworthiness. To induce more trust/trustworthiness in an experimental setting the authors, rather than coupling players randomly, did so according to the subjects' individual indicators of trust and trustworthiness[3] (sorted treatment). During the game subjects with a higher indicator of trust were paired with subjects with higher indicator of trust-worthiness. It is important to underline that subjects were not informed about the matching procedure; they thought that in every round they were playing with a randomly chosen person.

The analysis of the data showed that a 'behavioral clustering mechanism' boosts cooperative play significantly in comparison with the level reached by players randomly paired. Particularly at the end of the game, the fraction of subjects whose choice was to trust is much higher in the sorted treatment than in the random one. There is also considerable increase of trustworthiness among trustees during the game from 50 % to 83 %. The average efficiency score, i.e. how efficient play is with respect to the potential social benefit, is higher in the sorted treatment than in the random treatment. In the random treatment a player's initial level of trust seems to be overshadowed by the environment.

The Rigdon et al. (2007) experiment shows that a continuously enriched experience plays an important role while making a decision concerning trust. It is noteworthy that the emergence of trust and trustworthiness is a result of the interaction with 'cooperative' players, rather than a strategic choice.

Description of Experiment 2

McEvily et al. (2006) explored the factor of experience in the development of trust from a quite different angle. They examined how past experience of interaction with a member of a specific group will influence the outcome of future interactions with another member of that group. They hypothesized that there is a difference between trust placed in a particular individual and trust placed in someone who belongs to a collective entity. Although in both situations the interaction happens between two people in the second situation the factor is latent. This factor is a collective entity, which McEvily et al. (2006) describe as "an aggregate social system comprising a number of individuals."

The mechanism of trust in a collective entity is illustrated by the following example: let's imagine that you go to a physician about whom you do not know

---

[3] The indicator of trust depended on how many times the trustor showed trust and sent money to the trustee. The indicator of trustworthiness depended on how many times the trustee honored trust and returned money to trustor when they had a chance to do so.

anything. What could influence your decision whether or not to trust their diagnosis? Probably most of us would say that the crucial factor would be the behavior of the physician and their approach to making the diagnosis. However, one should also indicate a latent factor in this situation that could influence your decision – experience from all the interactions with other physicians in the past. So, if results of past interactions with the members of the same group do have a significant influence on our present and future attitudes toward individual members of that group, one should behave according to the rule: "if in the past my trust in a physician was honored, now I should trust the diagnosis."

In order to investigate this situation, McEvily et al. (2006) used the minimal group paradigm. Research on minimal groups (Tajfel et al. 1971) shows that people's behavior depends on their temporal identification: they may perform different actions when they act as individuals or as the members of a group. Moreover, people tend to change their behavior when they interact with a particular individual or with an individual who they see 'labeled' as a member of a particular group.

McEvily et al. (2006) designed an experiment in which they explored the influence of interaction with an anonymous member of a group on the judgment of trustworthiness of the whole group. They were predominantly interested in examining how trustors will respond to the outcome of the earlier round of a trust game when playing a new round.

They planned two treatments. In the control treatment subjects were divided randomly into two groups: trustors and trustees. They each played the trust game with anonymous players twice – in every round of the game they were coupled with a different counterpart. In the experimental treatment, on the other hand, subjects were assigned to the group of trustors or trustees on the ground of their earlier performance in a task in which they were asked to estimate the number of days it would rain the following year in San Francisco. The subjects who assessed that there would be many rainy days next year, were assigned to the group of trustors (Group A) and the rest to trustees (Group B). After the split of the subjects into two groups, trustors played the trust game with anonymous counterparts from Group B twice. The trustors were coupled with different counterparts in the two rounds. The division based on the irrelevant task for the estimation of trustworthiness of trustees was in accordance with the minimal group paradigm where the group identity can have a very weak form.

The analysis of experiments revealed that trustors in the experimental group reacted more strongly in the second game to experience in the first game, especially to a negative one, in which their trust was abused. Although the results were barely statistically significant, it is still striking that even in a very subtle group identity manipulation the betrayal of one member of a group can draw to decrease the expected trustworthiness of a whole group and therefore of individual members of the group.

The mechanism of trust in a collective entity can serve as a heuristic that people use once they are making a trust decision. It means that someone deciding whether to trust a particular person can look at them as the member of particular group (physicians, teachers, house wives, etc.) and unconsciously use the level of trust they have in the collective entity to estimate the person's trustworthiness.

These results show that our trust behavior in a given context is not shaped equally by all our experiences (e.g., mean outcome of all interactions) but rather by experience that is closest in nature to the situation being analyzed.

What can we say about the dynamics of trust and reciprocity once we refer to the results of experiments in which the role of experience was tested? Both mentioned experiments were conducted in a dynamical setting as they include changes in time. Looking at the results of the first experiment we can see that people start the game in different trust attractors: trust or no-trust. What happens next is the consolidation of their initial attractor due to the intentional clustering of players. We assume that the attractor deepens when the person adopts trust behaviors and their trust is reciprocated in consecutive time steps. The same happens with the opposite behavior of people with a deficit of trust and trustworthiness. Here a single act of trust, even if present, would most probably be abused by the trustee which would result in strengthening of no-trust attractor. We assume that at the end of the experiment the subjects' dominant attractors are very powerful and subjects would need to experience a dramatic change of setting to be pushed out of their attractors. But this was not verified so far in experiments.

However, what we can infer from the second experiment is that attractors of trust and no-trust can be assigned to groups and not generalized over the whole popula-tion. When we interact with individuals 'labeled' as physicians, bankers, policemen, etc., we develop assumptions regarding the whole group they belong to. The assumptions are based on the results of this encounter and they form the shape of attractors for further interactions. Based on the experiments of McEvily et al. we can hypothesize that it is particularly strong in the case of negative experience.

Role of Entanglement: Playing Both Roles in the Trust Game

People exhibit 'social preferences' while interacting with others. Instead of follow-ing the economic principle of maximizing their monetary payoff, they behave according to social norms such as reciprocity and fairness. A single accidental transaction with a stranger activates a different set of norms than a perspective of repeated interactions with the same counterpart. Moreover, one would expect to observe different sets of strategies when people anticipate having the same role in future transactions, compared to those who know that their roles are going to be more complicated: they will have to be trust-givers and trust-takers almost at the same time. Imagine three friends John, Steve and Carl. John is a student who is always short of money and wants to borrow from Carl who usually has some extra money. Carl is not really willing to lend to John because he does not know if John will return the money and if he will give him any extras for doing so, so he contemplates pretending not to have any extra cash. In this relation Carl is the person who is asked to lend. Would the behavior of Carl be different if he was hoping to get some extra money from Steve for the investment they were planning to do? Does the knowledge that a person can be in both roles change their behavior?

Burks et al. (2003) tested this situation with the help of the trust game. In their experimental setting each subject played twice, once as trustor and once as trustee. In each game players were paired with a different partner unknown to them. The difference between treatments was that in the first one trust-givers found out that they will also play the role of trustee only after having made a decision as trustor. In the second treatment, subjects knew that they would play both roles: trustor and trustee, at the beginning of the game, i.e., before they made a decision as trustor.

Burks et al. (2003) had two contradictory hypothesis. Firstly, they expected that playing both roles would increase both trust and trustworthiness because subjects could feel how it is to be in the other role. But secondly, they also expected that trust and trustworthiness could be decreased when subjects are asked to play both roles, especially, in the situation when they are asked to play the role of trustee before they are informed about the results on their first decision as a trustor. In this state they do not know if their trust was reciprocated and to what extent. This level of uncertainty might make participants feel less responsible for the well being of their partner and therefore more selfish. In such a situation the uncertainty overshadows the guilt.

The results revealed that playing both roles in the trust game confirm the second hypothesis: reduction of trust and overall reciprocity in the population. The data showed that when people are asked to play only one role in the trust game the mode for the trustor is to send almost all the money to the trustee, and for the trustee there are two modes: return 50–70 % of the received money but also return 0 %. In the situation when subjects are asked to play both roles, trustors send much more often smaller quantities of money to the trustee, and trustees return nothing to trustors more frequently.

This analysis suggests that in a dynamic setting in which subjects play both roles, they change their behavior in such a way that there is less trust and trustworthiness in the population.

However, before generalizing the results we need to be very careful with understanding the subjects' reasoning in the experimental setting. Maybe the experimental procedure was too complicated resulting in an increase of risk and decrease of willingness to trust, or maybe trustors felt that they will need to take the risk twice compared to their counterparts: firstly, because they do not know if they will be sent any money as trustee (they could get nothing), secondly, because they have to decide if they should send money as trustor (when sending they could receive nothing back).

Looking at this from the dynamic perspective we can say that the attractor can be shaped by a subjective interpretation of risk. In a situation when there are too many uncertainties, subjects can be pushed into a no-trust attractor even when their dominant mode is trust. This could be verified in an experiment in which the experimental setting would be adjusted by allowing the strength of subjects' dominant attractors to be measured.

Conclusion

In this chapter we pointed out that trust is a dynamic phenomena and showed that the trust game is a suitable tool for capturing this dynamics. Moreover, the dynamical interpretation of outcomes of trust game experiments enriches the understanding of trust and gives a new insight into various social and individual phenomena related to trust as well as reciprocity. Nonetheless, the findings of trust game experiments should be interpreted in light of the trust game's limitations. First, the research shows that the outcomes of the game can be very sensitive to its settings. Minor differences in the instructions and procedures of games may encourage subjects to follow very different norms, and use different strategies that result in dramatically diverse observed behavior (McCabe et al. 1998; Carpenter 2002). Moreover, naming a subject's counterpart as a 'partner' in one setting, and as an 'opponent' in the other, influences observed behavior and changes the outcomes of the game (Burnham et al. 2000).

Another and even more important limitation is that the trust game does not allow the testing of all trust behaviors but is constrained to those interactions that are based on monetary exchange. In other words, in the trust game trust comes down to money. And we can think of other situations in which trust plays a role but in which trust cannot be reduced to monetary value. For example, when we trust that our best friends will not be unfaithful to us or when we trust that our parents will take care of our children when we go to the movies. However, once we are able to measure the value of the subject of trust in monetary terms, the trust game seems to be a good choice for capturing the trust phenomena in laboratory setting.

References

Axelrod, R.: The Evolution of Cooperation. Basic Books, New York (1984)

Bachmann, R., Zaheer, A.: Handbook of Trust Research. Edward Elgar Publishing, Cheltenham (2006)

Baier, A.: Trust and antitrust. Ethics 96(2), 231–260 (1986)

Berg, J., Dickhaut, J., McCabe, K.: Trust, reciprocity, and social history. Game. Econ. Behav. 10, 122–142 (1995)

Burks, S.V., Carpenter, J.P., Verhoogen, E.: Playing both roles in the trust game. J. Econ. Behav. Organ. 51(2), 195–216 (2003)

Burnham, T., McCabe, K., Smith, V.S.: Friend-or-foe intentionality priming in an extensive form trust game. J. Econ. Behav. Organ. 43(1), 57–73 (2000)

Carpenter, J.: Information, fairness and reciprocity in the best shot game. Econ. Lett. 75, 243–248 (2002)

Dunn, J.R., Schweitzer, M.E.: Feeling and believing: the influence of emotion on trust. J. Pers. Soc. Psychol. 88(5), 736–748 (2005)

James, H.S.: The trust paradox: a survey of economic inquires into the nature of trust and trustworthiness. J. Econ. Behav. Organ. 47, 291–307 (2002)

Khaneman, D., Tversky, A.: Prospect theory: an analysis of decision under risk. Econometrica 47(2), 263–292 (1979)

Luhmann, N.: Vertrauen. Lucius & Lucius, Stuttgart (1968)

McAllister, D.: Affect- and cognition-based trust as foundations for interpersonal cooperation in organizations. Acad. Manage. J. **38**, 24–59 (1995)

McCabe, K., Smith, V., LePore, M.: Intentionality Detection and "Mindreading": Why Does Game Form Matter? Department of Economics, University of Arizona, mimeo (1998)

McCabe, K.A., Rigdon, M.L., Smith, V.L.: Positive reciprocity and intentions in trust games. J. Econ. Behav. Organ. **52**(2), 267–275 (2003). doi:10.1016/S0167-2681(03)00003-9

McEvily, B., Weber, R.A., Bicchieri, C., Ho, V.T.: Can groups be trusted? An experimental study of trust in collective entities. In: Bachmann, R., Zaheer, A. (eds.) Handbook of Trust Research. Edward Elgar Publishing, Cheltenham (2006)

McLeod, C., Trust. In: Zalta, E.N. (ed.) The Stanford Encyclopedia of Philosophy. Spring 2011 Edition, forthcoming. URL=<http://plato.stanford.edu/archives/spr2011/entries/trust/>

Potter, N.N.: How Can I Be Trusted? A Virtue Theory of Trustworthiness. Rowman & Littlefield, Lanham (2002)

Rigdon, M.L., McCabe, K.A., Smith, V.L.: Sustaining cooperation in trust games. Econ. J. **117**(522), 991–1007 (2007). doi:10.1111/j.1468-0297.2007.02075.x

Snijders, J.D.: Trust and Commitments. Thela Thesis, Amsterdam (1996)

Sterman, J.D.: Business Dynamics: Systems Thinking and Modeling for a Complex World. Irwin/ McGraw-Hill, Boston (2000)

Sztompka, P.: Zaufanie: Fundament Społeczeństwa. Wydawnictwo Znak, Kraków (2007)

Tajfel, H., Billig, M., Bundy, R.P., Flament, C.: Social categorisation and intergroup behaviour. J. Exp. Soc. Psychol. **1**, 149–177 (1971)

Uslaner, E.: Democracy and social capital. In: Warren, M. (ed.) Democracy and Trust. Cambridge University Press, Cambridge, UK (1999)

Chapter 12
Group as a Unit of Analysis

Karolina Lisiecka

Abstract A group can be viewed as an assembly of individual members but also as a collective entity with its own characteristics and dynamics. We argue that in order to understand connections between the individuals and the group-as-a-whole, one has to assume that groups are systems which can be analyzed at multiple, not just two levels. Below the level of the 'group-as-a-whole', there are many intermediate levels built one on top of one another. At each level of abstraction, a new organization emerges which has some causal power over the elements at lower levels. Higher level phenomena can be described as patterns built from elements at lower levels. In this chapter we review the existing theories that take a multilevel perspective on groups and conclude that this approach enables a well-founded empirical study of emergence of group as a collective entity.

Introduction

What is a social group? Can a group's properties be reduced to the attributes of individuals that form the group? When reflecting on their social experiences, people tend to focus on collective, rather than individual action. This is why we say "the orchestra played Mozart" instead of "the musicians played Mozart" or "the team scored a point" not "the player scored a point." Being holistic in our social perception, we even tend to bestow groups and teams with human qualities such as personality, memory, goals or free will (c.f. Allport 1933; Wilson 2005). The collective entities are seen as performing tasks, making decisions, or even experiencing emotions. But assigning human-like properties to groups is only a metaphor; groups are not animated beings.

K. Lisiecka
Department of Psychology, University of Warsaw, Stawki 5/7, Warsaw 00-183, Poland
e-mail: klisiecka@gmail.com

A. Nowak et al. (eds.), *Complex Human Dynamics*, Understanding Complex Systems,
DOI 10.1007/978-3-642-31436-0_12, © Springer-Verlag Berlin Heidelberg 2013

Two perspectives on group processes can be taken (Arrow et al. 2000; Burke 2003; Friedkin 2004; Moritz and Watson 1998). One focuses on the group as an emergent social system, with its own dynamics and global characteristics. The collective phenomena – norms, culture, or the structure of relations – ensure a group's sustainability and continuation. The other focuses on the study of individual members and their reactions: attitudes, emotions and behaviors. Further interaction at the individual level can be considered as a dynamic chain of members' actions, such as communicative acts, self-presentation behaviors (e.g. status demonstrations) and personal contributions to group goal attainment.

Mainstream social psychology has been primarily concerned with phenomena at the level of individuals, treating the micro perspective as the only legitimate experimental paradigm (Bar Tal 2006; Burke 2003). The fact that the group can be described both at the individual and the collective level is generally acknowledged in social psychology, even in popular textbooks on group dynamics (e.g. Brown 2000). It is also acknowledged that group-level properties emerge in the dynamic social interaction between individuals. However, what is not well understood, in our view, is how exactly this happens – how the group as a collective entity arises from a myriad of events occurring at the micro level.

In this chapter we will look into some of the existing accounts that describe connections between the individual and collective level of social reality. The theories presented below are in opposition to the traditional analytic – reductionist – methodology of scientific inquiry dominant in small group research today (Arrow et al. 2000; McGrath et al. 2000). As we will argue, a social group is a complex system which cannot be reduced to its basic elements – individual people and their intrapersonal processes. To understand how a collective entity is formed, one has to assume the existence of levels of description transcending the single individuals. These levels are essential for an accurate description of phenomena such as group process, shared reality or social structure. The emergent – holistic approach to groups that we present here helps, in our opinion, to better understand the causal relations and interactions between collective and individual phenomena.

The Group as a Gestalt

When a group is formed by people starting to interact, work together or talk, various fascinating phenomena begin to happen: a group hierarchy emerges, shared goals are created, people acquire unique roles and distribute responsibility in order to achieve those goals. A group spirit and ideology is created as the group maintains coherence and solidarity. These elusive phenomena start to affect members of the group so deeply that they may change their way of living in terms of habits, social identity and convictions. An individual becomes a representative of the group instead of being a 'free rider.'

Where do these phenomena exist? In the minds of the group members or in the eye of the observer? Can the existence of a collective level be proven empirically? One of the first social psychologists who admitted the real existence of 'social units,' such as group atmosphere, leadership or the concept of 'group' itself, was Kurt Lewin. At that time, the study of something without material existence was considered mystical and nonscientific. Lewin (1947) wrote:

> There is no more magic behind the fact that groups have properties of their own, which are different from the properties of their subgroups or their individual members, than behind the fact that molecules have properties which are different from the properties of the atoms or ions of which they are composed (Lewin 1997, p. 303).

Lewin's understanding of social phenomena developed from Gestalt psychology, influential in early twentieth century among German psychologists of mind (Koffka 1935; Köhler 1929; Wundt 1912). Gestalt means a 'figure' or a 'shape', referring to the discovery that human perception is holistic and creative and tends to organize details in the perceptual field into coherent wholes. With insights from Gestalt psychology, Lewin, in his research on 'group dynamics' (he was the originator of this term), called groups 'dynamical wholes,' i.e. systems of interdependent factors, in which change of any element results in the change of other elements (Lewin 1943).

Lewin's ambition was to apply the methodology of the exact sciences to psychology and the social sciences. He was the first to create formal definitions of abstract psychological constructs such as 'psychological position,' 'cognitive structure', 'goal,' 'conflict' and 'fear' (Lewin 1944). He aimed at breaking down these phenomena into elemental constructs such as 'field,' 'force,' 'locomotion,' 'strength' and other notions from field theory, that he adopted from physics. He was convinced that the language of field theory allowed for finding direct causal relationships between psychological constructs relevant in describing individual and group behavior.

According to Lewin's field theory, a group's social field is a Gestalt, i.e. a system consisting of individuals operating in the group's 'life space' (context). The field is defined as "the totality of coexisting facts which are conceived of as mutually interdependent" (Cartwright 1951, p. 240). At a given moment of time, various forces affect the social field of the group, invoking movement either toward a certain goal or state (helping forces) or away from this state (hindering forces). The change within the group life space is the resultant of the additive value of forces (or vectors) affecting the group from the inside (the pressures of members) and the outside (the pressures of the environment). Their combined effect results in the actual locomotion in the social field. Each force affecting the field can be represented along one dimension in phase space. A change in the value of one force changes the state of the whole system. One can trace the change by looking at its phase space representation.

One of the most appealing insights of Lewin's view on groups is that social events are not invariant in time. They undergo constant change, yet they can provide a stable environment for the individual. He described the social process vividly as "a process which, like a river, continuously changes its elements even if its velocity and direction remain the same. In other words, we refer to the characteristics of quasi-stationary processes" (Lewin 1997, p. 310). His explanation of the concept of quasi-stationary equilibrium can be directly related to the concept of attractor in complexity science (see this volume, Nowak et al., Chap. 1). For Lewin a process with quasi-stationary equilibrium fluctuates around a certain average value (equivalent to a point attractor). Its effect is restricted to a certain neighboring area within the force field (a basin of attraction). If the process reaches a level beyond the neighboring range, then it can loose its equilibrium and not return to the previous level (the boundary of a basin of attraction). The gradient of a neighboring state can be more or less steep (equivalent to the depth of the attractor). If it is steep, the fluctuations within the state of the social process will be smaller and the conduct of individual members of the group will be more similar.

In his work, Lewin has foreseen some statements of complex systems theory. Our conclusion from Lewin's depiction of group dynamics is that group-as-a-whole is not a static entity, it is an ongoing process. The group Gestalt emerges as a result of interdependency and co-occurrence of events within the force field. According to Lewin, to describe group-level phenomena it is not enough to define them, it is necessary to understand how they change in time in relation to other variables acting within the field.

Robert Bales, another great scientist studying small groups, also perceived them in terms of force fields. His methodology was different from Lewin's. His primary interest was a detailed description of the micro level – people's reactions and communication behaviors. For this purpose, he constructed two widely used coding systems, IPA – Interaction Process Analysis (Bales 1950, 1953; Bales and Strodtbeck 1951) and later, SYMLOG – System for Multiple Level Observation of Groups (Bales 1983, 1999).

In the IPA coding procedure, invented in the 1950s, each communication act in the observed interaction is assigned to a category by competent judges. Also, the sender and the receiver of each utterance is coded; this enables the activity of each member of the group to be tracked. Frequencies in each category are counted. Using IPA, Bales discovered the tendency for each member to dominate in a certain coding category (which was interpreted as the crystallization of group roles) as well as polarization of members around their level of activity or frequency of initiating interaction (Bales 1959). Bales described a typical communication struc-ture, in which members with a high status and attractiveness were privileged by being addressed more often than members of lower status and attractiveness. With IPA, Bales also described a typical development of events during a single meeting of task-oriented groups working on various problems (Bales and Strodtbeck 1951). He argued that a group solves a task by moving from an orientation phase, where the group concentrates on clarifying information, to an evaluation phase, involving assessment of information and ideas, and then to a control phase, when the final group decision is formed. He also discovered that the relative frequency of both positive and negative reactions increases over time.

Over a quarter of a century later, Bales and his colleagues developed the SYMLOG system (Bales 1983; Bales and Cohen 1979; Bales 1999). SYMLOG was both a coding system and a theory of groups. The idea behind SYMLOG referred to the Lewinian notion of a social field (Goździkiewicz and Bańka 2003). In SYMLOG, each behavioral act has to be considered in the wider context of the

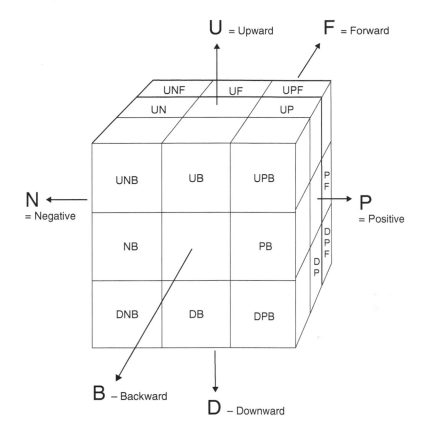

Fig. 12.1 Reprinted from: http://www.team-analysis.de/Symlog-methode.htm

totality of forces operating in the interactional field of the group. The field can be characterized by three dichotomous dimensions: right and left side (positivity versus negativity), upward and downward (dominance versus submission), forward and backward (goal-oriented versus emotional) – see Fig. 12.1. This last dimension was later changed to the acceptance versus rejection of task-orientation as specified by established authority.

SYMLOG can be used as a diagnostic tool for examining the configuration of forces affecting a group's interactional field and the relative position of members within this field (Bales and Cohen 1979). It consists of two elements: a coding system for competent judges (SYMLOG Interaction Scoring) and a short questionnaire for self-assessment by group members (SYMLOG Adjective Rating Form). Each member's average scores are represented visually on a so called 'field diagram'. The field diagram is similar to Lewin's phase space diagram – each axis represents one SYMLOG dimension. Just by looking at the positions of members in the SYMLOG space, one can deduce the role of this person in the group. For example, a person who scored high in dominance, positivity and task-orientation is clearly a group leader. If there is another person in the same area of the field, but higher on negativity or emotional dimension, clearly we might expect tension between those persons. SYMLOG can be used either for scientific purposes or as an informative feedback for the group members who might not be consciously aware of the process going on in their team.

A dynamic version of SYMLOG system was developed by Marcial Losada et al. (Losada and Markovitch 1990). Their GroupAnalyzer system enables the use of time series analysis in studying group level processes. In the GroupAnalyzer, each utterance, apart from being coded according to its sender, receiver and SYMLOG dimension, is assigned the time at which it took place. The timing of communication acts permits a dynamic animation of the group's interactional field to be made. This tool enabled Losada (1999) to make some discoveries about the dynamics governing the group field. For example, he compared task oriented groups who achieved success with the average and mediocre ones. The groups which were the most effective tended to behave in a more flexible manner – their members occupied more space in the field diagram and displayed a wider range of behaviors. Also, effective groups had a higher ratio of positive to negative utterances – above 2.9, on the so called Losada line (Fredrickson and Losada 2005).

Bales, Losada and their colleagues show how detailed observation of communication during group interaction enables the discovery of laws operating at the level of the group as a whole. Their work is an example of a multilevel theory of groups that allows for exact empirical study of the elusive group level phenomena. Yet, even though holistic in his thinking about groups, Bales used reductionists methods; he illustrated the group level as an accumulation of events occurring at the micro level. Losada (Losada 1999; Losada and Heaphy 2004) goes beyond pure aggregation of coded speech acts in group interaction. He describes correlations in time between members' communicative behaviors – the parameter he calls the connectivity of a team. This approach is qualitatively different from the linear reductionist approach that treats micro-events occurring in the group as independent. By looking at the product of the interaction between variables at a lower level, he is able to describe nonlinear effects and complex patterns emerging at the higher level – attractor dynamics of the group as a whole.

In the next section we will look into the theory of emergence, which claims that at higher levels of abstraction, new patterns come into being that are qualitatively different from the elements at the lower levels. Their properties cannot be fully derived from the properties inherent in elements of which they are composed.

The Group as an Emergent Entity

Emergence is the scientific stance which aims to investigate how patterns arise out of a multiplicity of relatively simple interactions (Nowak 2004; Goldstein 1999). An emergent, as formulated originally by Lewes (1875), is an effect that cannot be predicted from a full knowledge of its components. An emergent effect is

Fig. 12.2 'Levels of existence' (Source: Wikipedia: http://en. wikipedia.org/wiki/File: Levels_of_existence.svg)

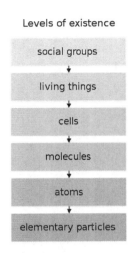

Levels of existence

social groups

living things

cells

molecules

atoms

elementary particles

considered to be not additive, not predictable from knowledge of its components, and not decomposable into those components (Lewes 1875; c.f. also Sawyer 2001; Goldstein 1999). In contrast to a resultant, an emergent property cannot be explained on the basis of conditions that led to it.

Emergentism assumes a multi-level structure of reality. At the same instant, a person is a part of society, an individual pursuing their own goals, an organism or a set of cells and atoms (Miller and Miller 1992). At each level of existence (c.f. Fig. 12.2), the behavior of the system is qualitatively different from the behavior of its elements. One cannot explain one level merely on the basis of another. A family or a social group is distinct from individual persons forming it, the mind is irreducible to interactions between neurons in the brain, and functions of the organism cannot be understood just by looking at the cells and molecules from which it is built. Sciences at each level of abstraction have to use a distinct set of scientific terms and laws and describe phenomena peculiar to that level.

Emergence has become extremely relevant in the twentieth century philosophy of science because it relates to the issue of the universality of scientific laws and translatability of notions between scientific domains. Emergentism offers a compromise between holistic and reductionist viewpoints in the philosophy of scientific inquiry (Wimsatt 2008; Simon 2008; Sawyer 2005). Emergentism disagrees with the reductionist perspective, which proposes that all concepts from a scientific domain at a higher level of organization can be inferred on the basis of explanations from domains of science at the lower levels of organization (see Fig. 12.2). Emergentists claim that at a higher level of organization, novel properties arise that are qualitatively different from their fundamental particles. The emergents are integrated wholes, therefore they cannot be fully understood and studied with reductionist methods. On the other hand, emergentism is not fully holistic because it admits that the only reality is physical matter. Emergent phenomena result from basic phenomena in a way that changes in the properties at a lower level inevitably lead to changes at a higher level (supervenience of levels).

Even though emergents arise out of fundamental entities, their properties have to be delineated at a different level of description.

Emergence in social systems has been one of the most hotly debated topics in social science of twentieth century (Sawyer 2005). Although many schools and approaches developed, traditionally relationships between just two levels have been studied: the basic level of the individual, where the unit of analysis is a single behavior, intention or cognition, and the emergent level of social structure, where the units of analysis are networks, roles, positions, distribution of resources and power (e.g. Lévi-Strauss 1963; Parsons 1937). In parallel to the theories stressing the fundamental role of social structure and its causal influence over the individual, a distinct approach developed that put emphasis on another level of social reality: communication and symbolic interaction between individuals (e.g. Blumer 1962, 1969; Turner 1968). From the perspective of symbolic interactionism, social structure is dynamically created in the process of symbolic exchange between participants of a social encounter. Communication is an interpretative process in which the meaning and importance of objects from the environment is negotiated. Social interaction is mediated by symbols which become institutionalized, giving rise both to the identity of an individual and to culture and collective norms. Thus, social encounter is primary both to human mind and to social structure (Mead 1934).

According to Sawyer (2001), contemporary sociology lacks a theory of emergence that would explain the mechanism of individual-to-structure transition through social interaction. He claims that even though emergentists reckon interaction as central, they have not supported their theories with the close empirical study of symbolic interaction. It is not clear how exactly crystalized social structure emerges from dynamic social interaction. A similar statement can be made about social psychology (Bar Tal 2006). Even though the reality of group-level phenomena, such as 'structure' or 'group thinking,' is acknowledged in the mainstream small-group psychology, no straightforward theory is provided as to how these phenomena emerge during group interaction.

Sawyer's own view on social emergence focuses on the description of 'micro-interactional mechanisms' that lead to the emergence of collective social phenomena (Sawyer 2005). In his Emergence Paradigm, he identifies two additional layers of social reality, localized in between the interaction level and social structure level: ephemeral and stable emergents. They describe the level of conversational encounter and emergence of collective behavior during social interaction and influence individuals in the sense that they "constrain the flow of interaction." Sawyer perceives social interaction as a continuing recreation of emergent patterns at the immanent layers of reality. He accepts that emergents, in some sense, exist objectively, as they cannot be explained on the basis of individuals' representation of them. They have causal power over social actors even if they have no knowledge of them.

Ephemeral emergents are what a group psychotherapist would call the group process, even though Sawyer does not relate them to this term directly. Instead, he refers to an interactional frame, defining "characters, motivations, relationships, plot events and sequence" in the social encounter. Ephemeral emergents are dynamic structures of conversation that emerge from collective action. Their existence is confined to a single encounter. They are not consciously accessible to individuals, yet they affect group members by constraining the flow of interaction. Because a single behavior in each turn acquires meaning only after its subsequent interpretation by the partners, the actual development of group process cannot be accredited simply to speakers intentions.

Stable emergents survive the single encounter; they are connected with more enduring group phenomena such as group culture and memory, social practices and conversational routines. They are processes responsible for group development and learning. Stable emergents are related to the sociological notion of 'collective behavior', phenomena such as mob actions, riots, mass delusions, crazes, fads, and fashions (Sawyer 2005, p. 211).

The idea of emergent communication structures helps to understand how durable social structure crystalizes during interaction. Sawyer claims that the jump from the individual to interaction and from interaction to social structure is simply too big. Transformation of the micro in the macro level is more gradual. The ongoing process of interaction involves the emergence of a whole array of more or less stable patterns.

A similar idea has been explored in some other interesting accounts. For example, Morgeson and Hofmann (1999) theorize on how, in organizations, collective structure emerges from collective action, and what a collective action really is. They propose the term 'event cycle' which represents a single individual action followed by a partner's responsive reaction. Collective action of the group is built out of an assembly of event cycles. Some micro patterns tend to be repeated and they gradually crystalize into collective structures which can be called the 'memory of the organization'. Salazar (1996), on the other hand, refers to Giddens' structuration theory in the development of group roles. He sees the establishment of social roles as a structuration process: production and reproduction of behavioral patterns. When entering a group, people first engage in role behaviors based on their previous experiences in social encounters and their own predispositions. They can also invent novel practices suitable to the particular social situation. The initial positioning in the 'role space' is either reinforced or rejected by the group. If it is reinforced, it is likely to become established as a situated social practice.

These theories on social emergence maintain that describing groups at only two levels (individual and systemic) is too simplified. Beneath the level of 'group-as-a-whole', there are lower-level sub-structures related to symbolic interaction and communication: collective memory, discourse patterns, roles and positions of members in interaction. Empirical study of connections between the micro and the macro patterns is not trivial and the introduction of intermediate levels can help to track emergence in groups. But finding better definitions of emergent levels of social reality is only the first step. The next is understanding the process that led to their emergence. A precise description of mechanisms behind forming the emergent patterns can be provided by the language of complex systems theory.

The Group as a Complex System

The notion of emergence is important in complexity theory. At the heart of complexity theory is how complex properties emerge in systems composed of elements interacting according to very simple rules (e.g. Goldstein 1999; Gilbert 2004; Bedau and Humphreys 2008). But complexity theory offers something more than theoretical accounts of emergence. It offers the possibility of studying the emergent processes empirically. This possibility is provided by an extensive usage of tools and methods from physics, mathematics, biology and computer sciences (see this volume, Nowak et al., Chap. 1).

Computer simulations of social systems usually deal with social phenomena at the level of a society. Nonetheless, the complexity approach has also been applied to the study of processes occurring in small groups. So far, Arrow et al. (2000) have offered the most elaborated theory of the group as a complex system. They postulate the existence of three levels at which causal dynamics in groups can be described. Local dynamics can be briefly described as typical daily routines ('activities') of a group. Those daily routines may be represented in a form of a coordination network with links connecting members via tasks they perform together and tools they share in order to fulfill these tasks. Once the coordination network is established, the group begins to function as a collective entity to which the rules of global dynamics can be applied. Global dynamics refers to variables reflecting the state of system as a whole. They are not 'simple aggregates of the local variables' but emergent features of the system. Global variables are in ongoing recursive relation to local variables, as they simultaneously arise from and constrain group operations at the local level. The third level, contextual dynamics, relates to external forces that affect both local and global dynamics. It reflects requirements of the environment in which a group is operating. Arrow et al. define the system complexity of a group as "the number and variety of identifiable regularities in the structure and behavior of the group, given a description of that group at a fixed level of detail." They argue that groups tend to increase in complexity over time. Group development can thus be perceived as gradual patterning in the temporal sequence of events occurring in the group. The emergence of regularity from seemingly random behaviors of group members may result from a natural press for coordination observed in social systems (Marsh et al. 2009). The pull to coordinate is regarded as an essential basis of social connectedness to others (e.g. Nowak and Vallacher 1998, 2003). Coordination in a social situation not only releases positive emotions and fosters cooperation (Wiltermuth and Heath 2009; Kulesza and Nowak 2007), it has a fundamental practical value. At each new level of analysis beyond individual behavior, new possibilities for action emerge (Marsh et al. 2009). A dyad can do more than a single person, a team more than a dyad and a multilevel organization more than a single workgroup. This observation can be summarized by the proverb: two heads are better than one.

In order to engage in joint action and tackle common goals and tasks, members have to react to each other. Mutual influence and interdependence between members enables cooperation and goal oriented co-action. When other individuals appear in the vicinity of an individual, their current state becomes constrained by the states of these individuals (Johnson 2006). Emergence of a coordination pattern is the emergence of correlation between states and actions of individual elements.

Coordination requires coupling of elements within a system. Basic elements of a social system – humans – are themselves complex systems that display rich dynamics. For the sake of simplicity, each person can be represented by an iterated nonlinear equation approximating the person's behavior in time (see this volume, Nowak et al., Chap. 1).

Gottman's general systems theory of marriage is also based on a set of coupled difference equations representing husband and wife (Gottman et al. 2002). The equations describe a sequence of verbal reactions of the conversing couple. At each time step, values of the equations denote positive or negative emotions expressed by the parties. One could expect that such a system would produce chaotic or random behavior, however it displays periods of stable behavior. These stable states describe distinct phases of conversation. In Gottman et al., the phases represent periods of positive and negative exchange between husband and wife. Positive and negative stable states are both possible for successful and unsuccessful marriages. The difference is that unsuccessful marriages display a higher emotional inertia and higher potential for the negative state which is sustained by mutual influence of the partners. A laboratory study of patterns in marital interactive behaviors allowed Gottman (1994) to predict with over 90 % accuracy whether a given couple will divorce or stay married.

Computer simulations allow for modeling of emergent phenomena at the level of societies. Yet, complexity based approaches can also be used in modeling the process established during a single social encounter. Coupling of interacting elements in the system produces sequences of predictable behaviors – patterns of coordination between partners. Interdependence and mutual adjustment of behavior is the mechanism underlying the emergence of a social unit. In the next section we give an example of an empirical study which captures a specific type of an emergent interactional pattern. Its methodology is embedded in the long tradition of social interaction observation (c.f. Gottman 1981; Bakeman and Gottman 1997; Gottman and Roy 1990).

An Example of Study of Emergent Coordination Process

In our study on group coordination, we examined the moment-to-moment relation between the structure of conversation and its meaning. We expected that during interaction in a discussion group, momentary patterns of coordination will emerge which can be related to the state in which a group is at a given moment of time. We focused on a group conversation about controversial issues where two clear opposing stands can be taken, either for or against some cause. We hypothesized that such a situation may evoke moments of heated discussion between members, if not an open conflict between opposing parties. When a group starts a controversial topic, the course of discussion may change radically to a state of disagreement. An interaction that was up until this point friendly and agreeable, may transform itself into an exchange of dominance, negativity and advocacy for own opinions.

In our study, we video-taped 8 four-person groups talking about abortion. We analyzed the video recordings to track the sequential order of turn-taking during each conversation. Also, the content of each utterance was evaluated by independent judges. We used 5 seven-point Likert scales: dominance–submission, positivity–negativity, inquiry–advocacy, agreement–disagreement and prochoice–prolife. Also, before each conversation we asked participants about their opinions on abortion. After each conversation we measured their satisfaction with the interaction process.

Coordination patterns can be traced in the sequence of members taking the floor in a conversation. We asked ourselves if there is any specific pattern of turn-taking that is characteristic of a state of disagreement in a group discussing a controversial topic. In our analysis we focused on short, four-element sequences of turn-taking. There are only five possible sequences of this kind, involving two, three or all four members of the group (ABAB, ABAC, ABCA, ABCB or ABCD). Out of all these patterns, the obvious pattern of disagreement between two contradictory stands is the ABAB pattern. It is the sequence in which two people speak alternately, in relation to each other.

In a typical pattern of disagreement in a dyad, two people escalate their arguments round after round and finally end up in a quarrel. The same can be expected in the ABAB pattern in a group interaction. If the ABAB pattern is indeed related to moments of disagreement in a group, it should occur at the same time as escalation patterns in the semantic content of the utterances: stronger opinions stated, stronger advocacy of those opinions, higher levels of dominance, disagreement and negativity.

We expected that on some occasions the ABAB-type exchanges would involve more than four subsequent utterances. Sometimes two people would talk alternately for an extended period of time, e.g. ABABA-BABA. We expected that these more lengthy ABAB exchanges would mark the moments of the most heated discussion, when a conversation turns to a particularly controversial subject. We supposed that the prolonged disagreement in a group, marked by the lengthy ABAB exchanges, should involve more negative statements, more direct expression of members' opinions and stronger advocacy of their positions. In a group interaction, reactions of other members have to be taken into account. They can either join in the quarrel or cool it down by trying to find common grounds or changing the subject. If they join in, they can try to support the arguments of one party disregarding the ones of another. Thus, the pattern of disagreement can be sustained, even when the players are changed.

At first, our analysis concentrated on distinguishing the ABAB patterns in the conversations studied. An utterance was coded as belonging to an ABAB pattern whenever it was a part of a sequential exchange between two speakers. We compared the semantic content of ABAB and non-ABAB utterances. We found significant differences on several semantic dimensions. Utterances belonging to the ABAB patterns were more dominant, more negative, showed higher levels of advocacy of own opinions and more disagreement towards statements of other group members. Also, the opinions (both prolife and prochoice) stated in the ABAB utterances were stronger.

We also wanted to know whether the more prolonged ABAB sequences were associated with the moments of the highest disagreement. In the next step of the analysis, the length of the ABAB pattern was coded in the sample. Two ABAB sequences were treated as the same ABAB exchange when they took place immediately one after another, even if the speakers were changed. We correlated the lengths of ABAB sequences with variables describing their semantic content. It turned out that the longer the ABAB sequence, the higher were the levels of dominance, negativity, advocacy and disagreement conveyed by the utterances in the sequence. Even though significant, Spearman's rank correlation coefficients were very low for the raw data (from $\rho = -0.1$, $p < 0.001$ for agreement–disagreement dimension to $\rho = 0.156$, $p < 0.001$ for opinion strength). To obtain a clearer picture of the relationship between the variables, we aggregated the data with the length of ABAB sequence as the breaking variable. For the aggregated data, we found substantial correlations between the length of ABAB-type exchanges and three semantic variables: positivity–negativity ($\rho = -0.520$, $p = 0.033$), inquiry–advocacy ($\rho = -0.640$, $p = 0.006$) and opinion strength ($\rho = 0.515$, $p = 0.035$). The more lengthy the ABAB exchange, the less positive were the statements of which it was composed; these statements also conveyed stronger opinions (both prolife and prochoice) and more advocacy instead of inquiry.

Finally, we asked ourselves if the members who were involved in particularly lengthy ABAB exchanges during the conversations differed somehow from other members of the groups. We expected that these persons would differ in their opinion on abortion from the rest of the group, which would explain why they engage in disagreement exchanges during the conversations. We also expected that their assessment of the entire conversation would be less positive, as being in disagreement with the group would probably be difficult and unpleasant for them. In our analysis we used three measures obtained from the questionnaires filled in by the group members: level of satisfaction from interaction process, strength of opinion on abortion (either prolife or prochoice) and dissimilarity between the person's initial opinion on abortion and the mean opinion in the group. These three measures were correlated with the maximum length of the ABAB exchanges a particular person participated in during a conversation. We found significant correlations between the maximum length of the ABAB exchanges and all of these measures: satisfaction ($\rho = -0.430$, $p = 0.014$), opinion strength ($\rho = 0.361$, $p = 0.042$) and members' dissimilarity from the group ($\rho = 0.355$, $p = 0.046$).

Our analysis of group discussions about abortion shows that phases of disagreement during a conversation can be identified as momentary patterns of coordination in sequence of turn-taking. The occurrence of the ABAB pattern – the exchange between two members speaking alternately – marks the moments of confrontation between the two opposing stances: prolife versus prochoice. This disagreement state is characterized by higher levels of dominance, negativity, disagreement, strength of the opinions uttered, and advocacy of those opinions. Our results also show that disagreement states which are longer or sustained by other members of the group are more intense. What is interesting, participation in particularly lengthy ABAB-type sequences tends to affect members' assessments of their entire conversation.

The study shows supervenience of two levels in the group interaction: the level of the semantic content of utterances exchanged by the members and the level of the momentary interaction frames which capture the sequence of communicative behaviors. The meaning of a given utterance, whether it is positive or negative, agreeing or disagreeing, is contingent on the intentions of an individual member of the group, the speaker. The ABAB-type communication structure is a feature of the interaction-as-a-whole because it transcends the individual member. The pattern of turn-taking cannot be accredited to each individual speaker alone as none of them can determine how their current utterance will be interpreted and who will be the next speaker (c.f. Sawyer's description of ephemeral emergents). Rather, it is a manifestation of a higher-order pattern of disagreement that, within a given instant of time, transforms behavior of the group as a whole.

The group disagreement can be perceived as a set of processes occurring simultaneously at different levels of abstraction. These processes run in parallel – they take place in the same time; they are interdependent but a causal direction between them cannot be indicated. Is the ABAB-type pattern the cause or the effect of increased intensity of the utterances? It is more likely that both processes trigger one another in an ongoing feedback loop between the emergent level (the coordination pattern characteristic of a disagreement) and its base (the meaning of each utterance).

Summary

In the approaching twilight of the traditional reductionist approach to science, a modern scientist has to face new challenges which result from a growing sophistication of knowledge about the surrounding world. As stated by Johnson (2010):

> The modern world is characterized by problems that involve systems with social and physical subsystems. They are entangled systems of systems with multilevel dynamics. There is no methodology able to combine the partial micro-, meso- and macro-theories that focus on subsystems into a coherent representation of the dynamics of the whole. (Johnson 2010, p. 1)

Despite the devotion of social science to the micro approach (Bar Tal 2006), an increasing number of social scientists urge their colleagues to develop cross-level explanations of social phenomena (Bar Tal 2006; Chan 1998; Coleman 1990; Rousseau and House 1994; Sidanius and Pratto 1999) and some authors respond to this request (e.g. Friedkin 2004; Erez and Gati 2004; Morgeson and Hofmann 1999; Moritz and Watson 1998). At present we may be witnessing a growing zeitgeist for multilevel models applied to social science.

A social scientist who studies group processes has to find a way to deal with the complexity of their subject. The emergent-holistic approach to groups that we presented above does not make things easier. In this perspective, a group as a collective entity is an intricate, hierarchical structure of relationships between processes at different levels of abstraction, where each emergent level is a pattern composed of elements at the lower level. The emergent patterns are qualitatively different from the elements at the lower levels. As exemplified by our study on emergent coordination process in group interaction, the link between the emergent and the base level can be observed empirically as the co-occurrence of events at both levels of abstraction.

References

Allport, F.H.: Institutional Behavior: Essays Toward a Re-Interpreting of Contemporary Social Organization. University of North Carolina Press, Chapel Hill (1933)

Arrow, H., McGrath, J.E., Berdahl, J.L.: Small Groups as Complex Systems. Formation, Coordination, Development, and Adaptation. Sage Publications, London (2000)

Bakeman, R., Gottman, J.: Observing Interaction: An Introduction to Sequential Analysis, 2nd edn. Cambridge University Press, New York (1997)

Bales, R.F.: Interaction Process Analysis: A Method for the Study of Small Groups. Addison-Wesley, Cambridge, MA (1950)

Bales, R.F.: The equilibrium problem in small groups. In: Parsons, T., Bales, R.F., Shils, E.A. (eds.) Working Papers in the Theory of Action, pp. 111–161. Free Press, Glencoe (1953)

Bales, R.F.: Task roles and social roles in problem solving groups. In: Maccoby, E.E., Newcomb, T.M., Hartley, E.L. (eds.) Readings in Social Psychology, 3rd edn. Methuen, London (1959)

Bales, R.F.: SYMLOG: a practical approach to the study of groups. In: Blumberg, H.H., Hare, A.P., Kent, V., Davies, M. (eds.) Small Groups and Social Interaction, vol. 2, pp. 499–523. Wiley, New York (1983)

Bales, R.F.: Social Interaction Systems: Theory and Measurement. Transaction, New Brunswick (1999)

Bales, R.F., Cohen, S.P.: SYMLOG: A System for the Multiple Level Observation of Groups. Free Press, New York (1979)

Bales, R.F., Strodtbeck, F.L.: Phases in group problem-solving. J. Abnorm. Soc. Psychol. **46**, 485–495 (1951)

Bar Tal, D.: Bridging between micro and macro perspectives in social psychology. In: Van Lange, P.M. (ed.) Bridging Social Psychology: Benefits of Transdisciplinary Approaches, chapter 50, pp. 341–346. Lawrence Erlbaum Associates Publishers, Mahwah (2006)

Bedau, M., Humphreys, P. (eds.): Emergence: Contemporary Readings in Philosophy and Science. MIT Press, Cambridge, MA (2008)

Blumer, H.: Society as symbolic interaction. In: Rose, A.M. (ed.) Human Behavior and Social Process: An Interactionist Approach. Houghton-Mifflin, Boston (1962). Reprinted in Blumer (1969)

Blumer, H.: Symbolic Interactionism: Perspective and Method. University of California Press, Berkeley (1969)

Brown, R.J.: Group Processes: Dynamics Within and Between Groups, 2nd edn. Blackwell, Oxford (2000)

Burke, P.J.: Interaction in small groups. In: DeLamater, J. (ed.) Handbook of Social Psychology, pp. 363–388. Kluwer-Plenum, New York (2003)

Cartwright, D. (ed.): Field Theory in Social Science: Selected Theoretical Papers by Kurt Lewin. Harper & Row, New York (1951)

Chan, D.: Functional relations among constructs in the same content domain at different levels of analysis: a typology of composition models. J. Appl. Psychol. **83**, 234–246 (1998)

Coleman, J.S.: Foundations of Social Theory. Harvard University Press, Cambridge, MA (1990)

Erez, M., Gati, E.: A dynamic, multi-level model of culture: from the micro level of the individual to the macro level of a global culture. Appl. Psychol. Int. Rev. **53**(4), 583–598 (2004). doi:10.1111/j.1464-0597

Fredrickson, B.L., Losada, M.F.: Positive affect and the complex dynamics of human flourishing. Am. Psychol. **60**(7), 678–686 (2005)

Friedkin, N.E.: Social cohesion. Annu. Rev. Sociol. **30**(1), 409–425 (2004)

Gilbert, N.: Agent-based social simulation: dealing with complexity. Downloaded 15.12.2010, from http://www.agsm.edu.au/bobm/teaching/SimSS/ABSS-dealingwithcomplexity-1-1.pdf (2004)

Goldstein, J.: Emergence as a construct: history and issues. Emergence **1**, 49–72 (1999)

Gottman, J.M.: Time Series Analysis: A Comprehensive Introduction for Social Scientists. Cambridge University Press, Cambridge (1981)

Gottman, J.M.: What Predicts Divorce? Lawrence Erlbaum Associates, Hillsdale (1994)

Gottman, J.M., Roy, A.K.: Sequential Analysis: A Guide for Behavioral Researchers. Cambridge University Press, New York (1990)

Gottman, J.M., Swanson, C., Swanson, K.R.: A General systems theory of marriage: nonlinear difference equation modeling of marital interactions. Pers.Soc. Psychol. Rev. **6**(4), 326–340 (2002)

Goździkiewicz, A., Bańka, A.: SYMLOG jako metoda pomiaru i predykcji sukcesu w zarządzaniu. In: Witkowski, S. (ed.) Psychologiczne wyznaczniki sukcesu w zarządzaniu, Prace Psychologiczne 6, 215–225 (2003)

Johnson, J.: Can complexity help us better understand risk? Risk Manage. **8**(4), 227–267 (2006)

Johnson, J.: The future of the social sciences and humanities in the science of complex systems. Innov. Eur. J. Soc. Sci. **23**(2), 115–134 (2010)

Koffka, K.: Principles of Gestalt Psychology. Harcourt, Brace & World, New York (1935)

Köhler, W.: Gestalt Psychology. Horace Liveright, New York (1929)

Kulesza, W., Nowak, A.: Dlaczego Ciebie lubię? Bo się koordynujemy. In: Winkowska-Nowak, K., Rychwalska, A. (eds.) Modelowanie matematyczne i symulacje komputerowe w naukach społecznych. Wydawnictwo SWPS Academica, Warszawa (2007)

Lévi-Strauss, C.: Structural Anthropology. Basic Books, New York (1963)

Lewes, G.H.: Problems of Life and Mind, Series 1, vol. 2. Trubner, London (1875)

Lewin, G. (ed.): Resolving Social Conflicts & Field Theory in Social Science. American Psychological Association, Washington, DC (1997)

Lewin, K.: Defining the 'Field at a given time'. Psychol. Rev. **50**(3), 292–310 (1943). Reprinted in Lewin, G. (ed.): Resolving Social Conflicts and Field Theory in Social Science, pp. 200–211. American Psychological Association, Washington, DC (1997)

Lewin, K.: Constructs in psychology and psychological ecology. University of Iowa Studies in Child Welfare, **20**, 1–29 (1944). Reprinted under the title "Constructs in field theory," in Lewin, G. (ed.): Resolving Social Conflicts and Field Theory in Social Science, pp. 191–199. American Psychological Association, Washington, DC (1997)

Lewin, K.: Frontiers in group dynamics 1. Hum. Relat. 1, 5–41 (1947). Reprinted in Lewin G. (ed.): Resolving Social Conflicts and Field Theory in Social Science, pp. 301–366. American Psychological Association, Washington, DC (1997)

Losada, M.: The complex dynamics of high performance teams. Math. Comput. Model. **30**(9–10), 179–192 (1999)

Losada, M., Heaphy, E.: The role of positivity and connectivity in the performance of business teams. Am. Behav. Sci. **47**(6), 740–765 (2004)

Losada, M., Markovitch, S.: Group analyzer: a system for dynamic analysis of group interaction. In: Proceedings of the 23rd Annual Hawaii International Conference on System Sciences, pp. 101–110. IEEE Computer Society Press, Washington, DC (1990)

Marsh, K.L., Richardson, M.J., Schmidt, R.C.: Social connection through joint action and interpersonal coordination. Top. Cogn. Sci. **1**(2), 320–339 (2009)

McGrath, J.E., Arrow, H., Berdahl, J.L.: The study of groups: past, present, and future. Pers. Soc. Psychol. Rev. **4**, 95–105 (2000)

Mead, G.H.: Mind, Self, and Society. University of Chicago Press, Chicago (1934)

Miller, J.G., Miller, J.L.: Cybernetics, general systems theory, and living systems theory. In: Levine, R.L., Fitzgerald, H.E. (eds.) Analysis of Dynamic Psychological Systems, vol. 1, pp. 9–34. Plenum, New York (1992)

Morgeson, F.P., Hofmann, D.A.: The structure and function of collective constructs: implications for multilevel research and theory development. Acad. Manage. Rev. **24**, 249–265 (1999)

Moritz, S.E., Watson, C.B.: Levels of analysis issues in group psychology: using efficacy as an example of a multilevel model. Group Dyn. Theory Res. Pract. **2**(4), 285–298 (1998)

Nowak, A.: Dynamical minimalism: why less is more in psychology. Pers. Soc. Psychol. Rev. **8**(2), 183–192 (2004)

Nowak, A., Vallacher, R.R.: Dynamical Social Psychology. Guilford Press, New York (1998)

Nowak, A., Vallacher, R.R.: Synchronization dynamics in close relationships: coupled logistic maps as a model for interpersonal phenomena. In: Klonowski, W. (ed.) From Quanta to Societies. Frontiers on Nonlinear Dynamics, vol. 2, pp. 165–180. Pabst Science Publishers, Berlin (2003)

Parsons, T.: The Structure of Social Action. Free Press, New York (1937)

Rousseau, D.M., House, R.J.: Meso organizational behavior: avoiding three fundamental biases. In: Cooper, C.L., Rousseau, D.M. (eds.) Trends in Organizational Behavior, vol. 1, pp. 13–30. Wiley, London (1994)

Salazar, A.J.: An analysis of the development and evolution of roles in the small group. Small Group Res. **27**, 475 (1996)

Sawyer, R.K.: Emergence in sociology: contemporary philosophy of mind and some implications for sociological theory. Am. J. Sociol. **107**(3), 551 (2001)

Sawyer, R.K.: Social Emergence: Societies as Complex Systems. Cambridge University Press, Cambridge (2005)

Sidanius, J., Pratto, F.: Social Dominance: An Intergroup Theory of Social Hierarchy and Oppression. Cambridge University Press, New York (1999)

Simon, H.A.: Alternative views of complexity. In: Bedau, M.A., Humphreys, P. (eds.) Emergence: Contemporary Readings in Philosophy and Science, pp. 249–258. MIT Press, Cambridge, MA (2008)

Turner, R.: Role: sociological aspects. In: Sills, D. (ed.) International Encyclopedia of the Social Sciences. Macmillan, New York (1968)

Wilson, R.A.: Collective memory, group minds, and the extended mind thesis. Cogn. Process. **6**(4), 227–236 (2005)

Wiltermuth, S., Heath, C.: Synchrony and cooperation. Psychol. Sci. **20**(1), 1–5 (2009)

Wimsatt, W.C.: Aggregativity: reductive heuristics for finding emergence. In: Bedau, M.A., Humphreys, P. (eds.) Emergence: Contemporary Readings in Philosophy and Science, pp. 99–110. MIT Press, Cambridge, MA (2008)

Wundt, W.: An Introduction to Psychology. Macmillan, New York (1912)

Chapter 13
Conflict as an Attractor: A Dynamical Systems Perspective on the Dynamics of Conflict

Lan Bui-Wrzosinska

Abstract This chapter presents, from a dynamical systems perspective, theoretical and empirical work on the processes of conflict at the interpersonal, social, and inter-group level, as well as practical implications from such works. The research discussed suggests that it is particularly useful to conceptualize ongoing destructive conflicts as resulting from strong attractors: a specific form of self-organization of multiple elements consisting of mental and social systems associated with the conflict. The dynamic properties of conflict in family dyads, marriage and group dynamics are illustrated and quantified using tools derived from complexity science, with particular emphasis on identifying patterns of self-organization leading to destructive social relations. The chapter begins by demonstrating the relevance of the concept of attractors for a description of the dynamics of conflict processes, at different levels of social reality. Next, a series of novel hypotheses, lines of research and interesting results in the domain of interpersonal conflict emerging from this approach are presented, with special emphasis on methods adopted in various areas of psychology for the measurement and quantification of conflict attractor patterns. The chapter concludes with theoretical and practical implications of the presented framework for conflict resolution and transformation, with a detailed description of an attractor software – a visualization tool used in various conflict resolution settings – that has proven particularly effective in supporting lasting, sustainable solutions in complex conflict negotiations.

Understanding, predicting and managing conflicts are arguably among the most important challenges facing mankind. On the one hand, disputes and disagreements, being inherently human activities, are essential to the construction of a shared reality, problem solving, democratic governance, and adaptability. On the other hand, beside their invaluable potential, conflicts between people, groups and states

L. Bui-Wrzosinska
Warsaw School of Social Science and Humanities, Chodakowska 19/31, 03815 Warsaw, Poland
e-mail: lwrzosin@gmail.com

A. Nowak et al. (eds.), *Complex Human Dynamics*, Understanding Complex Systems,
DOI 10.1007/978-3-642-31436-0_13, © Springer-Verlag Berlin Heidelberg 2013

profoundly damage relationships, and the well-being and development in families, societies and nations. The high cost of conflicts, involving loss and harm inflicted by marital dispute, gang fights, intra-organizational conflicts, political or ethnic struggles, especially when manifested through violence, provide a strong rationale for prevention and termination of destructive, unproductive cycles of conflict escalation. Surprisingly, people engage in maladaptive, violent and destructive conflicts, even when the costs of such processes considerably exceed their expected benefits. Effectively, conflict – as a basic and largely adaptive feature of social life – often degenerates into a persistent pattern of behavior that brings out the worst in human nature (Vallacher et al. 2010). The frequently irrational character of conflict dynamics, as well as its complex nature involving both potential for growth and for destruction, presents a substantial challenge to the social sciences.

While providing insight into human motives in conflict situations, neither utility theories nor existing perspectives on human social and individualistic motivation offer a framework sufficient for the understanding of people's consistent engagement in destructive and unproductive conflicts. Recently, a scientific paradigm widely used in other areas of science – the dynamical systems approach (DST) – has been applied to the study of conflict (Coleman et al. 2006; Coleman et al. 2007; Coleman et al. 2010; Nowak et al. 2007; Vallacher et al. 2010). This theoretical advance has opened new avenues for the study of complex systems of conflict, bringing novel hypotheses and conceptual tools to the field. Here, we propose that a key to the understanding of basic human and social drive toward destructive conflicts is the DST notion of attractor. An attractor 'attracts' the system's behavior: if conflict becomes an attractor for the system, even very different features, such as individual motives, orientations, and positions tend to evolve toward the subset of conflict states. This concept emphasizes the self-organization of the system's elements into global patterns that, once formed, resist perturbation and other sources of change (cf. Holland 1995; Kelso 1995; Schuster 1984; Strogatz 2003), and provides a strong rationale to use attractor dynamics to study people's engagement in destructive conflicts.

We start this chapter by presenting how the concept of attractors contributes to a description of the dynamics of conflict at different levels of social reality. Next, we describe a series of novel hypotheses, lines of research and interesting results emerging from this approach in the domain of interpersonal conflict. We conclude with the theoretical and practical implications of the presented framework for conflict resolution and transformation.

Conflict as an Attractor

In generic terms, an attractor refers to a subset of potential states or patterns of change to which a system's behavior converges over time (see this volume, Nowak et al., Chap. 1). When an attractor governs the system's dynamics, the system becomes resistant to perturbing influences, both from outside and from inside; the system also becomes resistant to individual tendencies of its constituent elements that would otherwise move it to a different state or pattern of changes. This approach allows us to identify several rules which a conflict system governed by attractor dynamics would follow.

First, it implies that if there is an attractor for social or individual states defined by conflict, then escalation into conflict is much easier than de-escalation. This dynamical rule has not been explicitly stated in the conflict literature; on the contrary, conflict escalation is presented rather as a progression, an uphill rise in intensity, while de-escalation is presented as a decrease in intensity. It has been hypothesized that such widespread one-dimensional mapping of conflict dynamics along a single intensity axis may be based on incorrect assumptions and on inadequate mental models of the dynamical properties and forces acting on conflict systems (see: Bartoli et al. 2011).

Second, the attractor concept gives us insight about the nature of changes in conflict dynamics. DST emphasizes that in response to perturbations, systems with attractors undergo phase transitions – discontinuous, abrupt changes. The magnitude and nature of the transition from one state to the other, however, is not proportional to the perturbing stimulus, but represents a radical, qualitative shift rather than an incremental progression. Properties of such scenarios are of particular relevance for conflict de-escalation and practical applications: the hysteresis effect described in the catastrophe theory (Thom 1975), for instance, describes how small perturbations will result in the system returning to its original state, but further changes of the control parameters may result in the system abruptly moving toward a different set of stable states: threshold effects are to be expected in the system's trajectory. In dynamical social psychology terms, similar dynamics have been understood as catastrophic scenarios of change (Tesser and Achee 1994). Such scenarios may explain how crossing certain thresholds in conflict escalation leads to catastrophic, irreversible changes at the level of key social-psychological parameters, and may undermine the potential for further de-escalation.

Third, the attractor concept provides a platform for integration of dynamic data from different time scales, and levels of social reality (Nowak 2004). People's behavior, thoughts, or emotions at the micro level can be combined into patterns of interactions, and linked to higher-level emerging properties of the social, interpersonal or dyadic system. In a state of conflict, for example, the self-organization of individuals' thoughts, emotions, and behaviors may lead to the emergence of group level attitudes, prejudice and hostility. In this context a temporal trajectory of different states of the system may be analyzed, leading to accurate understanding and prediction of further conflict patterns.

Parameters of Conflict Dynamics

The main challenge in studying conflicts from a dynamical systems perspective lies in the accurate identification of the key parameters of the system. A dynamical system can generally be conceptualized as the state of its elements at a given time. A system's behavior is a sequence of such states. However, in order to describe this behavior, we need to identify the state space (the ensemble of possible states) within which the state trajectory will evolve, as well as the key variables and parameters affecting the evolving characteristics of the system (Nowak 2004). Variables identified in the conflict literature include static measures, such as conflict styles (Rahim 1983), individual differences, cooperation/competition (Deutsch 1973), or utility functions (Ross 1973; Myerson 1997). These measures, however, still need to be translated into dynamic variables, and parameters affecting the systems' evolution over time. In the next sections, several lines of research demonstrating the operationalization of such dynamic variables affecting systems of conflict are presented.

DST Research

Social scientists have developed various methodologies, and are pursuing various lines of study, in which attractors are identified in social, group and dyadic interactions. However, little of the research using the DST approach has focused on conflict. In this chapter we describe the methods used in these studies, from different fields of psychological research, and summarize their results. Most of this research has been conducted on interpersonal conflicts; from the DST perspective, conflicts at the social and inter-group level still remain relatively under investigated. We begin by briefly presenting research strategies adopted in developmental psychology, with emphasis on results shedding light on the dynamics of conflicts in parent–child interactions, involving adolescent-parent settings. Then, we introduce Gottman's approach to marital interactions, and demonstrate the value of the DST predictive models in assessing potential for divorce and distress in relationships. Then, we present recent advances and work in progress toward research strategies centrally concerned with conflict dynamics at different levels of social reality.

Parent–Child Interaction

Dynamic systems researchers have addressed two key areas of parent–child interaction: development of communication and socio-emotional development (Lewis et al. 1999). Both of these processes are self-organizing; that is, the development of communicative and emotional patterns are hypothesized to emerge and solidify through repeated parent–child interaction experiences over time (Granic and Lamey 2002).

For example, Granic and Dishion (2003) have identified attractors for deviant talk in adolescent friendships (measured as a tendency to increasingly repeat patterns of deviant talk exchange and return to such patterns after perturbation); these were predictors of further delinquency, pathological social development, poor social relations, drug abuse and escalation of interpersonal conflict. Another classic example of a conflict attractor emerging from developmental studies is the coercive cycle described by Patterson and colleagues (Patterson et al. 1998; Reid and Patterson 1989). In this cycle, an aggressive child's coercive behavior is reinforced through repeated instances of ranting on the part of the parent, resistance from the child, and the eventual withdrawal of the parent. Of course, all families experience some conflict, but it has been shown that families of aggressive children more often engage in extended and persistent exchanges of coercive behavior (Patterson et al. 1998). Such a result shifts the focus of therapeutic attention from a psycho-pathological diagnosis of maladapted children, to the family systems that lead to the emergence of destructive patterns of behavior.

It has been shown that attractors for hostile interaction emerge when there is a decrease in the flexibility of the behavioral repertoire on the part of both children and parents. It is the inability to flexibly exit those destructive interactions that leads to psycho-pathology (Gottman and Notarius 2000). These lines of study suggest that it is rigidity in social interactions, not the valence of the exchange itself, that is the best predictor of conflict and destructive interactions. Thus, it is not only the content but also the structure of parent–child behavior over time that accounts for the child's social adjustment. Most investigations have focused on the negative content of social interactions, although rigidity in the structure of interaction may apply to both positive and negative content. The problem arises when people are 'stuck' in a narrow set of possible states. That rigidity in mutual positive states is problematic seems to be counterintuitive. However, parents in well-adjusted families typically engage in regulatory mechanisms in response to extended periods of high positive arousal in children, alternating states rather than remaining perpetually positive.

Lewis et al. (1999) developed a graphical approach that utilizes observational data to quantify dynamic variables that define the state space for a dyadic system. This technique has been adapted for the further study of parent–child interactions (Granic et al. 2003; Granic and Lamey 2002). With the Gridware method, the sequence of behavioral states of two people interacting over time is plotted on a grid representing all possible behavioral combinations of the dyad. The parent's coded behavior is plotted on the x-axis and the child's behavior on the y-axis. The resulting trajectory represents a series of moves from one dyadic state to another over the course of an interaction.

Using this method, Granic and Lamey (2002) have been studying dysfunctional parent–child dyads, where the child has been diagnosed either with externalizing or mixed (both externalizing and internalizing) behavioral problems. Interactions of parent–child dyads during common problem-solving tasks were observed, and their reaction to intervention aimed at provoking more adaptive behavior was recorded. The observed sequence of behavioral states was plotted on a grid representing four possible states for each person (positive, neutral, negative, hostile), making 16 possible states in all.

The results revealed the same pattern of interaction during the problem-solving task for both groups of dyads (with externalizing and mixed children), where the parent would remain permissive (positive/neutral state) and the child would stay aggressive (hostile/negative). Prompting the dyads to move toward positive interactions, introduced no change when the child was diagnosed with externalizing problem behavior. However, in dyads in which the child was diagnosed with both externalizing and internalizing behavior problems, this manipulation introduced a catastrophic shift from permissive to escalating negativity and increasingly hostile interaction between the child and the parent. The first group had only one attractor for the relationship, where the parent would reinforce the child's aggressive behavior by being permissive, whereas the second group would exhibit two attractors: permissiveness of the parent in spite of hostile behavior of the child, or a recurring cycle of mutual hostility and conflict escalation. In these groups the dyad moved from one attractor (permissive parent/aggressive child) to another (aggressive parent/aggressive child). In both groups, the existing attractors for the parent–child interaction were so strong that even external perturbations could not dislodge the dyad from such patterns. It is clear that an attractor for a state of positive-positive interactions was lacking. In DST terms, we would hypothesize that a repellor exists in the state space for the relationship – a repelling state in this region of the state space (Keating and Miller 2002).

Results from developmental studies indicate that not only existing attractors, but also repellors states that are impossible in a given interaction, are critical for the prediction of trajectories of conflict and informed intervention strategies. Additionally, the data suggest that a conflict may escalate even in response to interventions aimed at resolving maladaptive behavior. Consequently, by imposing pressure on the system's elements, one needs to be aware of the natural tendencies of the system toward catastrophic shifts conditioned by the attractor structure. From this perspective, a traditional anticipation of classic cause and effect reactions of the system to intervention intended at reducing conflict is naive and may prove inaccurate.

The development studies suggest important hypotheses yet to be tested experimentally. Here, we report some preliminary research aimed at translating observation grids into experimental settings. Using the provocation scenarios methodology (Bui-Wrzosinska et al. 2009), the reaction of interpersonal systems to a series of perturbations, as well as change of people's behavioral repertoire in conflict, has been assessed. This method assesses attractor dynamics by actively perturbing the system through a sequence of conflict provocation stimuli. The psychological repertoire is measured on a scale of 30 possible behaviors ranging from 'talking it over', through 'yelling at him/her', to 'killing him/her'; participants are asked which behaviors from the scale are possible, and which cease to be possible as the conflict situation evolves. Changes of one participant's psychological repertoire are measured as a reaction to another party's (X) conflict provocation steps described in a series of vignettes (e.g. "imagine X humiliates you in front of your coworkers"). The stimuli have the aim of gradually destabilizing the relationship between the two parties through the use of hypothetical scenarios (vignettes); how the relationship stabilizes after perturbation is registered. One way to formally portray and systematize results collected through such manipulation is to map the results on a grid, where the manipulation steps are represented on the x-axis, and the reactions of the participants are represented on the y-axis; the dynamic properties of the resulting conflict trajectories are further described using attractor mapping. It is hypothesized that if a single-point attractor exists, the system will always return to the same state after perturbation. In the case of multiple fixed-points, small perturbations will result in the system returning to its original state, but further changes of the control parameter may result in the system moving toward a different stable state, and thus threshold effects, such as dramatic escalation or outburst of violent behavior are to be expected in the response patterns.

A preliminary study investigating the effect of the relational closeness between two parties on conflict escalation and de-escalation trajectories has been conducted using this method. The most obvious difference between close and more distant relationships is the valence of the affective relation between the participants, being more positive for close relationships. Another difference, less trivial and critical from the point of view of this study, is that close relationships exert pressure toward consistency in attitudes, emotions, and in the overall patterns of interactions between the parties, restraining the scope of the behavioral repertoire to positive behaviors; a more distant relationship induces fewer constraints on the interaction. The study revealed that closeness characterizing friendship between two people, is related to an initial narrow behavioral repertoire, involving a set of unequivocally positive possible behaviors. However, in response to stronger perturbations induced through conflict provocations, such dyads react abruptly, with qualitative shifts in the relationship, while dyads having more distant relations are characterized by gradual escalation trajectories. Close relationships lead to trajectories with a major shift after a series of responses, where, despite contentious behaviors from the other party (from situations, where your friend does not answer your phone calls or emails, to situations, where he criticizes your work in front of other people), the responses initially remain at a very low level of hostility (the behavioral repertoire initially comprised: "listening to him/her", "turning it agreeably into humor", etc...). However, when a threshold of provocation was crossed, the psychological repertoire underwent a qualitative shift, and, in response to a single perturbation step moved from a conciliatory repertoire straight to a sequence of responses characterized by extremely high levels of hostility and open aggression ("hurt him/her as much as possible"). Empirical results show a nonlinear progression of responses from one stable state of positive relations toward another stable state with extremely aggressive relationships. On the other hand, dyads having more distant relationships have more gradual escalation patterns, where mid-range levels of provocation triggered an intermediate responses repertoire. Figure 13.1 illustrates the general patterns of escalation for the close and distant relationship conditions.

Note that a single static measure at a given moment in time would not predict these paradoxical effects of closeness condition on the system's dynamics: from a static point of view, close friends are expected to uphold a stable, positive relationship (Bukowski et al. 1998). The DST perspective supports this, but only when the levels of provocation are low.

Close / Distant relationships

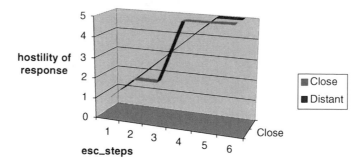

Fig. 13.1 Trajectories of escalation for close and distant relationships

Additionally, results from further lines of research show that the trajectory of escalation is critical for later de-escalation: if intermediary states are omitted, and conflict moves from low to high levels of hostility abruptly, then the newly formed attractor for the relationship is less amenable to de-escalation attempts. Results from these studies are a direct validation of the hysteresis effect described in catastrophe theory (Thom 1975; Poston and Stewart 1996), and support the catastrophic scenarios of change described by Tesser and Achee (1994).

Although still in its fancy, a line of research on the dynamics of change at the level of behavioral possibilities in response to provocation may shed light on the paradoxical effects of conflict dynamics on social relationships. For example, the unexpected, very rapid dynamics of escalation of conflict that ultimately led to unthinkable acts of cruelty and violence between neighbors that had been coexisting peacefully for decades, such as those observed in former Yugoslavia or Rwanda, may have a coherent explanation in the dynamical systems theory approach. Escalation, from this perspective: (1) occurs not only at the level of observable behaviors, but also at the level of behaviors that start to be possible within a given interaction, be it between two people, groups, or nations, and behaviors that vanish from the range of possibilities. (2) Contrary to what the name itself suggests, conflict escalation does not always occur in a step-by-step fashion; it can be non-linear, moving abruptly from low to high intensity without necessarily going through intermediary steps. (3) The 'reverse engineering' of escalation processes is not symmetric: the amount of energy needed to 'undo' escalating steps is greater than the energy needed to escalate a conflict. (4) Moderate levels of conflict seem to be possible only if no strong pressure is exerted on the system, even if the pressure forces the system into exclusively positive states, so the dynamics of conflict may not stabilize at mid-range levels of intensity. As noted earlier, and in accordance with the observations of family dyads, unequivocally positive interactions, with strong pressure against negative feedback may not always be adaptive and may paradoxically lead to destructive conflict patterns.

Marital Conflict

Besides their diagnostic and empirical value, DST models also hold predictive potential. Gottman and his collaborators extensively observed distressed and non-distressed couple dynamics in their laboratory. Gottman and Levenson (2002) explored the strength of couple's conflict attractor by assessing their ability to recover from a conflict discussion by imposing a positive discussion upon the couple several minutes later. This method was aimed at perturbing the system's dynamics while it was positioned within a conflict region of the state space. Using this DST strategy, the strength of the basin of attraction for conflict patterns was assessed for the couple. These observations predicted divorce with 92.7 % accuracy. In DST terms, couples who were able to stabilize on the positive conversation shortly after a conflict discussion proved to have a strong attractor for positive interactions, while couples showing disgust, anger, or runaway responses during the imposed positive conversation, demonstrated the effect of a continuous pulling force toward the conflict attractor and a week attractor for positive interactions. It is noteworthy that these indicators were predictive of people's later divorce with 97.7 % accuracy.

Another study by Gotman and Levenson (2002), conducted over a period of 14 years, confirmed that couples that were negative during conflict early in their marriages were likely to divorce earlier in their marriages. In DST terms, such couples had a strong attractor for conflict, and were positioned within this attractor, and thus the short-term trajectory of their marriage appeared to be governed by this attractor (the amount of energy needed to dislodge them from the current attractor state was considerably undermining the possibility for the relationship to last). On the other hand, couples who were disengaged or had no positive emotions in both benign and conflict discussions was an indicator of later, rather than earlier divorce. In DST terms, this would correspond to a lack of positive attractor for the relationship, therefore such a system landscape is also predictive of divorce, but in the long-run, as it does not necessarily have a strong attractor for conflict. Buehlman et al (1992) also found strong predictors of divorce in couples' narratives of their past history. From this perspective, past trajectories reveal the structure of the attractor landscape and can predict future interactions. A discriminant function analysis showed that the oral history variables could predict divorce or marital stability with approximately 94 % accuracy.

Gottman, in his studies of couples' conflict engagements (Gottman 1991), also advanced a balance theory of marriage based on which combinations of styles (validating, avoiding, volatile, hostile, and hostile-detached) were likely to self-organize into coherent, stable states, and which were unstable and thus likely to fail in the long-run. From his data, partner attractors, forces or perturbations capable of altering patterns of dyadic interaction, alternative attractors (steady states following perturbation), and coupling functions (the positive or negative influence that the partner's state has on the other) were identified for both types of couples. Perturbation was not an external influence but an intrinsic variable as partners influenced each other and so perturbed each other from their individual attractors.

Gottman et al. (1999) extended this research by analyzing newly married couples, searching for thresholds of negativity and positivity for each spouse. Consistent with dynamic systems theory, the authors hypothesized that there was a threshold at which the negativity or positivity expressed by one spouse began to perturb the other spouse's affect and behavior. The authors again found that spouses had distinct initial attractors and also attractors formed through the couple's interactions. Interestingly, the initial attractors of couples who ended up divorced or unhappily married were significantly more negative than the attractors of couples who were still happily married 3–6 years later. Probably the most interesting parameter used in Gottman's studies was the coupling of physiological measures. It was already known that couples with a higher physiological linkage (partners would mutually influence and reinforce each other's state), appeared more predictable in their interaction trajectories and were more prone to conflict escalation (Levenson and Gottman 1983).

Practical Implications

Dynamical Systems Theory creates new possibilities for the study of conflict; it suggests innovative methods, theories and models that have strong predictive and empirical significance for the field. Another vital feature of DST models is their heuristic value. The concepts of attractors and of state space trajectories support the development of complex mental representations, such as field of forces and system energy, to represent the dynamics of abstract phenomena. These notions seem conceptually abstract, but are key to the understanding of conflict dynamics, and thus of the substantive effects on people's responses, strategies and behavior in conflict becomes tangible.

An innovative software tool based on attractors (Nowak et al. 2010), was first developed as a method for comprehending and addressing chronic patterns of destructive conflict and violence in New York City public schools. Later it has become a platform for teaching participants in multi-stakeholder negotiations in various protracted social conflicts. The main element of the teaching platform is a computer simulation of conflict attractors which permits the participants to visualize the dynamics of their conflict and to interact appropriately as these unfold over time.

From the dynamical systems perspective, enduring conflicts can be characterized as the consequence of attractors within a given social system. If the attractor for destructive conflict is strong (wide and deep), then the system will move toward a state of conflict and, even when disturbed by outside influences, will return to this state. However, to fully understand conflict dynamics in terms of attractor changes, one needs more than a description of the current attractor for the destructive conflict. The strength of the alternative attractor for peaceful and positive social interactions is also critical, particularly in the case of conflict transformation strategies. A conflict system's dynamics can be caught and controlled by a destructive attractor, where the attractor's tendencies can be seen in the processes specific to that conflict (such as overt hostility and violence). An attractor can also remain dormant, and may be able to capture the state of the system in the future. To sum up, a conflict system can be characterized on three dimensions: (a) its current (visible) state, (b) its ability to generate positive interactions, and (c) its ability to generate negative interactions.

Describing conflict on these three dimensions presents challenges to those unfamiliar with the dynamical systems approach, especially so for those initially unfamiliar with the concept of attractors. Moreover, each and every factor influencing a social system characterized by conflict may also have a different effect on each of the three dimensions, thus rendering systematic analyses of the relationships between the conflict system's elements almost impossible. For instance, enhanced police presence in an inter-group community conflict may momentarily decrease the state of violence in the community, but it may also decrease the strength of the attractor for future positive interactions between the parties and, analogically, increase the strength of the attractor for future negative interactions. Understanding multiple consequences of an action can be difficult for a person involved in a dynamic conflict situation.

The attractor software provides a visualization tool. It prompts the user to specify both the key factors influencing the conflict and any actions that can be undertaken as well as to estimate the consequences of these actions with respect to three types of outcomes:

1. their influence on the current state of the conflict;
2. their influence on the potential for future conflict or negative interactions and
3. their influence on the potential for positive social interactions and, ultimately, sustainable peace.

The user, by evaluating each factor, estimates the strength and the direction of the influence of each factor on the whole system. The software merely visualizes the understanding of the user, as a tool for systematically describing what parties and interveners have identified, based on the user's own expertise and experience with a case. The software itself does not estimate the importance of each factor nor its influences. It is up to the user to:

1. specify the case and the social relations to be analyzed (e.g., a marriage, an ethnic conflict);
2. create the list of factors that are likely to influence the nature of the current and future relationship and
3. evaluate the importance and strength of each factor for the system, and the direction of its influence along the three dimensions considered. (See Fig. 13.2; the list of input factors is accompanied by an illustration of the attractor landscape and the momentary state represented by the ball moving towards the valley.)

The program provides a visual depiction of two possible attractors for the relationship: a positive attractor indicating stabilization of benign or favorable

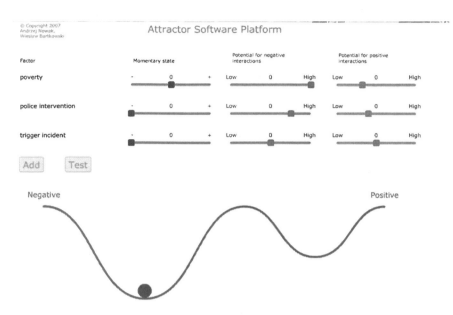

Fig. 13.2 The interface of the attractor software with no landscape

attitudes and positive actions, and a negative attractor indicating stabilization of malignant or unfavorable attitudes and negative or violent actions. How the factors introduced by the user affect the overt thoughts and behaviors in the relationship will thus depend on the respective strength and depth of the positive and negative attractors. If the current state of the relationship (e.g., good versus bad feelings) is within the basin of a strong attractor, then this state is unlikely to change, despite the introduction of factors relevant to change. On the other hand, if the current state of the relationship is outside the basin of the attractor, then the relationship may display a qualitative change (e.g., from positive to negative) with the addition of a single, seemingly unimportant factor.

The program relies both on the user's experience with the conflict at issue as well as expertise in the area of the conflict. Specifically, it assumes that the user possesses some knowledge about the factors relevant to the thoughts, feelings, and behaviors that characterize the analyzed relationship and also that they can specify their sense of the relative importance of these factors. For example, someone who works with high school gangs may be able to identify the various events that affect each gang's feelings, thoughts, and behaviors. However, such analyses gain more depth and higher validity if they directly involve various members of the groups involved. The users then type in the label for a factor and use a slider bar to specify its overall importance in affecting the relationships within the system.

Despite the user's expertise and insight, the influence of the specific factors that have been introduced into the visualization software is not obvious. As noted earlier, a minor provocation can push two groups into open warfare, while a major change in their conditions might have little effect on relations between the two groups. Attractor dynamics helps explain such nonlinearity between influencing factors and the observed state of the relationship. This can be seen when the attractors characterizing the relationship are described so that the resulting dynamics can then be visualized by the user. It stems from the fact that a momentary change in thought or behavior in response to a specific factor does not necessarily affect the long-term features of the relationship. However, the strong potentials of these systems in the form of latent attractors, sometimes become manifest as qualitative shifts in the relationship (e.g., radical shifts transform peaceful relations into conflict or violence into peace). If the state of a person or group is currently controlled by an attractor, then even strong forces that seem capable of changing thought and behavior may be countered by the attracting tendency of the prevailing thoughts and behaviors. But if a person's or group's state is currently at the edge of the basin of one attractor, then even a minor force might be sufficient to move that person's or group's thoughts and actions toward a completely different attractor.

In addition to using their knowledge and insight to specify relevant factors and their respective importance, the program's user may also specify the nature of their impact. Some factors can have both a positive and negative impact. In a marital relationship, for example, raising children can strengthen the bond between the partners, but it can also produce considerable stress and thus challenge the relationship. To capture a factor's potential for both positive and negative effects, the user employs separate slider bars to indicate how much the factor in question promotes positive thoughts and behaviors, and how much this factor boosts negative thoughts and behaviors in the relationship.

We note that characteristics of relationships do not always act in opposition; they often prove to be orthogonal. Consequently the potential for positive interactions can grow or decrease somewhat independently from the potential for negative interactions. For example, fostering social contact between conflicted groups can increase the potential for both positive and negative interactions. The program also allows the user to specify, on a slider bar, the degree to which a factor contributes to eruptions of violence. Again, the user's expertise and insights are critical here. In ethnic relations, for example, income disparity may be an important factor in the long run, but it is unlikely to directly spark an episode of violence on a particular day. An act of humiliation, in contrast, may well catalyze momentary violence in such a relationship. A separate slider bar allows the user to specify the impact of each factor on the immediate versus long-term reactions of the system.

Finally, the program allows the user to directly reconfigure the attractor landscape. The preset configuration of positive and negative attractors may not fully capture the user's knowledge and insight. It may happen, for example, that the user finds the positive attractor to be relatively weak but to possesses a wide basin of attraction, whereas the negative attractor is quite strong but has a narrow basin of attraction. By changing the attractor landscape, the user can observe whether the relevant factors and their specific effects on positivity, negativity, and momentary violence begin to play a larger or smaller role in defining the overall quality of the relationship. The software can thus be employed in different ways and will achieve different results.

The program's benefits include managing complexity, untangling the long-term versus short-term consequences of conditions and actions, and understanding that the same action can have conflicting effects on the positive and negative aspects of the interaction. In more general terms, the software allows its users to directly experience dynamic concepts and tools, and thus fosters their ability to understand conflicts. Here we outline some of the software's possible uses and benefits:

1. The interactive nature of the program enables students, negotiators, and third parties to use their knowledge of and insights about a conflict to see how factors affect the momentary and long-term state of the relationship between the conflicting parties. With each additional factor, the state of the relationship at that moment is changed, but whether this change affects the long-term relationship will depend on the attractor landscape. This should sensitize the users to the distinction between interventions that have immediate but not long-lasting effects and interventions whose effects may not be immediately apparent but that change the attractor landscape and raise new possibilities for relationships.

2. Social science theory and data can be used to identify relevant factors in particular situations, to specify the overall importance of these factors, and to define the impact of these factors on positivity, negativity, and momentary violence. Consequently, the program can be used to test the assumptions of existing social science models, as well as to identify the factors that should receive attention in real-world contexts. On the other hand, if the role of these factors has been unequivocally established in research and real-world contexts, one can modify the attractor landscape to adapt the results of the program to such findings. This 'reverse engineering' can help researchers and practitioners identify the manifest and dormant attractors in interpersonal and inter-group relations. Identifying the inactive attractors is particularly important because they represent possible relationship states that might otherwise go unrecognized by the parties involved, but that could motivate an intervention.

3. Users can work with the software in small groups, allowing them to share insights and together identify relevant factors, specify the effects of these factors, and observe how the relationship responds in both the short and long term. For example, rather than arguing about the probable effects of various interventions, the users can test their respective assumptions and intuitions, and in this way try to reach a common understanding with agreed-upon strategies for conflict resolution.

4. The software can also be used as a platform for resolving conflicts between representatives of conflicting parties. The parties to a conflict often perceive the world in different terms, and this lack of a shared reality may intensify the conflict's intractability. By working with this software in a collaborative venture, the conflicting parties can discover which factors constitute the foundations of the conflict. More importantly, an initiative of this kind could promote an agreed-upon mode of intervention for resolving the conflict.

Conclusion

The dynamical perspective has proven to have both an integrative and a heuristic value for the study of conflict dynamics. Results from a wide area of research have brought consistent support for the idea that systems of destructive conflicts exhibit attractor properties. Although the dynamical systems approach to conflict analysis, research, intervention and training is at an early stage, it nonetheless holds great promise for conceptualizing, comprehending and working with conflict dynamics in all types of social relations. Particularly, a systematic study of the behavioral repertoire of people, groups and nations in a DST context shows predictive and diagnostic potential. From this perspective, nonlinear patterns of interactions in conflict, as well as the remarkably stable character of some instances of destructive, unproductive social relations and seemingly paradoxical social processes find coherent explanation in self-organizing mechanisms. The attractor software tool and negotiation training pedagogy described in this chapter is just one of the many practical initiatives that may emerge from working with these new theories and technologies.

References

Bartoli, A., Nowak, A., Bui-Wrzosinska, L.: Mental models in the visualization of conflict escalation and entrapment: biases and alternatives (June 25, 2011). In: IACM 24th Annual Conference Paper. SSRN: http://ssrn.com/paper=1872605 (2011)

Bukowski, W. M., Newcomb, A. F., Hartup, W. W.: The company they keep: Friendships in childhood and adolescence. Cambridge University Press Cambridge (1998)

Buehlman, K., Gottman, J.M., Katz, L.: How a couple views their past predicts their future: predicting divorce from an oral history interview. J. Fam. Psychol. $5(3-4)$, 295–318 (1992)

Bui-Wrzosinska, L., Gelfand, M., Nowak, A., Severance, L., Strawinska, U., Cichocka, A., Formanowicz, M.: A dynamical tool to study the cultural context of conflict escalation. In: Modeling Intercultural Collaboration and Negotiation (MICON) Workshop (July 13, 2009), Passadena (2009)

Coleman, P.T., Bui-Wrzosinska, L., Vallacher, R., Nowak, A.: Approaching protracted conflicts as dynamical systems: guidelines and methods for intervention. In: Schneider, A., Honeyman, C. (eds.) The Negotiator's Fieldbook, pp. 61–74. American Bar Association Books, Chicago (2006)

Coleman, P.T., Vallacher, R.R., Nowak, A., Bui-Wrzosinska, L.: Intractable conflict as an attractor: presenting a model of conflict, escalation, and intractability. Am. Behav. Sci. **50**, 1454–1475 (2007). doi:10.1177/0002764207302463

Coleman, P.T., Vallacher, R., Nowak, A., Bui-Wrzosinska, L., Bartoli, A.: Navigating the landscape of conflict: applications of dynamical systems theory to protracted social conflict. In: Ropers, N. (ed.) Systemic Thinking and Conflict Transformation. Berghof Foundation for Peace Support, Berlin (2010)

Deutsch, M.: The Resolution of Conflict: Constructive and Destructive Processes. Yale University Press, New Haven (1973)

Gottman, J.M.: Predicting the longitudinal course of marriages. J. Marital Fam. Ther. **17**(1), 3–7 (1991)

Gottman, J., Levenson, R.W.: A two-factor model for predicting when a couple will divorce: exploratory analyses using 14-year longitudinal data. Fam. Process **41**(1), 83–96 (2002)

Gottman, J., Notarius, C.I.: Decade review: observing marital interaction. J. Marriage Fam. **62**(4), 927–947 (2000)

Gottman, J.M., Swanson, C., Murray, J.: The mathematics of marital conflict: dynamic mathematical nonlinear modeling of newlywed marital interaction. J. Fam. Psychol. **13**, 3–19 (1999)

Granic, I., Dishion, T.J.: Deviant talk in adolescent friendships: a step toward measuring a pathogenic attractor process. Soc. Dev. **12**, 314–334 (2003)

Granic, I., Lamey, A.V.: Combining dynamic systems and multivariate analyses to compare the mother-child interactions of externalizing subtypes. J. Abnorm. Child Psychol. **30**, 265–283 (2002)

Granic, I., Hollenstein, T., Dishion, T.J., Patterson, G.R.: Longitudinal analysis of flexibility and reorganization in early adolescence: a dynamic systems study of family interaction. Dev. Psychol. **39**, 606–617 (2003)

Holland, J.H.: Emergence: From Chaos to Order. Addison-Wesley, Reading (1995)

Keating, D.P., Miller, F.K.: The dynamics of emotional development: models, metaphors, and methods. In: Lewis, M.D., Granic, I. (eds.) Emotion, Development, and Self-Organization: Dynamic Systems Approaches to Emotional Development, pp. 373–392. Cambridge University Press, New York (2002)

Kelso, J.A.S.: Dynamic Patterns: The Self-Organization of Brain and Behavior. MIT Press, Cambridge, MA (1995)

Levenson, R.W., Gottman, J.M.: Marital interaction: physiological linkage and affective exchange. J. Pers. Soc. Psychol. **45**, 587–597 (1983)

Lewis, M.D., Lamey, A.V., Douglas, L.: A new dynamic systems method for the analysis of early socioemotional development. Dev. Sci. **2**, 457–475 (1999)

Myerson, R.B.: Game Theory: Analysis of Conflict. Harvard University Press, Boston (1997)

Nowak, A.: Dynamical minimalism: why less is more in psychology. Pers. Soc. Psychol. Rev. **8**, 183–192 (2004). doi:10.1207/s15327957pspr0802_12

Nowak, A., Vallacher, R.R., Bui-Wrzosinska, L., Coleman, P.T.: Attracted to conflict: a dynamical perspective on malignant social relations. In: Golec, A., Skarzynska, K. (eds.) Understanding Social Change: Political Psychology in Poland. Nova Science Publishers Ltd., Hauppauge (2007)

Nowak, A., Bui-Wrzosinska, L., Coleman, P.T., Vallacher, R.R., Bartkowski, W., Jochemczyk, L.: Seeking sustainable solutions: using an attractor simulation platform for teaching multi-stakeholder negotiation in complex cases. Negot. J. **26**, 49–68 (2010). doi:10.1111/j.1571-9979.2009.00253.x

Patterson, G. R., Reid, J. B., & Dishion, T. J.: Antisocial boys. Blackwell Publishing, Oxford (1998)

Rahim, M.A.: A measure of styles of handling interpersonal conflict. Acad. Manage. J. **26**(2), 368–376 (1983)

Reid, J. B., & Patterson, G. R.: The development of antisocial behaviour patterns in childhood and adolescence. Eur. J. Personality, **3**(2), 107–119 (1989)

Ross, S.A.: The economic theory of agency: the principal's problem. Am. Econ. Rev. **63**(2), 134–139 (1973)

Schuster, T.: Deterministic Chaos: An Introduction. Monographie, vol. 248. Seiten, Physik Verlag, Weinheim (1984)

Strogatz, S. H.: Sync: The emerging science of spontaneous order. Hyperion, New York NY (2003)

Tesser, A., Achee, J.: Aggression, love, conformity and other social psychological catastrophes. In: Vallacher, R., Nowak, A. (eds.) Dynamical Systems in Social Psychology, pp. 95–109. Academic, San Diego (1994)

Thom, R.: Structural Stability and Morphogenesis: An Outline of a General Theory of Models. W. A. Benjamin, Inc., Reading (1975)

Vallacher, R.R., Coleman, P.T., Nowak, A., Bui-Wrzosinska, L.: Rethinking intractable conflict: the perspective of dynamical systems. Am. Psychol. **65**(4), 262–278 (2010)

Printed by Publishers' Graphics LLC
DBT130422.11.38.138